COMPLEX NUMBERS

$$\sqrt{3+4i}$$

$$\frac{5+10i}{4-3i}$$

$$\cosh\left(\frac{\pi}{6}i\right)$$

$$\ln i$$

Chris McMullen, Ph.D.

Complex Numbers Essentials Math Workbook with Answers
Chris McMullen, Ph.D.

Copyright © 2024 Chris McMullen, Ph.D.
www.improveyourmathfluency.com
www.monkeyphysicsblog.wordpress.com
www.chrismcmullen.com

All rights are reserved. However, educators or parents who purchase one copy of this workbook (or who borrow one physical copy from a library) may make and distribute photocopies of selected pages for instructional (non-commercial) purposes for their own students or immediate family members only.

Zishka Publishing
ISBN: 978-1-941691-47-2

Mathematics > Precalculus
Mathematics > Complex Numbers

Contents

Introduction	iv
1 Imaginary Numbers	5
2 Complex Numbers	14
3 Complex Conjugates	17
4 Complex Arithmetic	20
5 The Complex Plane	27
6 The Modulus	31
7 Polar Form	35
8 Real and Imaginary Parts	39
9 Quadratic Roots	41
10 De Moivre's Theorem	49
11 Euler's Formula	57
12 Complex Trigonometry	63
13 Roots of Complex Numbers	75
14 Roots of Unity	89
15 Roots of Polynomials	94
16 Properties of Complex Numbers	111
17 Applications	123
Appendix: Cubic and Quartic Formulas	133
Answer Key	147
Index	220

Introduction

This book is designed to help students learn or review essential concepts relating to complex numbers and to provide practice applying the concepts to solve a variety of problems. Sample topics include:

- What is an imaginary number? What is a complex number?
- Why does multiplying by the complex conjugate result in a real number?
- How do you add, subtract, multiply, or divide complex numbers?
- What is the complex plane? How do you plot a number in the complex plane?
- Is there anything tricky about taking the square root of a negative number?
- How does the modulus square relate to the Pythagorean theorem?
- When is it useful to work with the polar form of a complex number?
- Does the cube root of 8 really have three possible answers?
- How do complex numbers relate to trigonometry?
- What are hyperbolic cosine and hyperbolic sine?
- Are hyperbolic cosine and sine related to the ordinary trig functions?
- Can you find the square root of i? If so, how?
- What is Euler's formula? What is De Moivre's theorem?
- Does $\sqrt{i}$ have both real and imaginary parts? Why isn't it purely imaginary?
- Is it possible to take the cosine of an angle and get a number larger than one?
- What happens if you take the logarithm of a negative value?
- If the coefficients of a polynomial are real, can any of its roots be complex?
- How do you find the roots of a polynomial?
- What is the fundamental theorem of algebra? Does it involve complex numbers?
- Which properties do complex numbers have?
- If you raise i to the power of i, what do you get?
- In what way are complex exponents and logarithms periodic?
- What paradox did Thomas Clausen encounter in 1827?
- Are there any common applications of complex numbers?

1 Imaginary Numbers

A <u>real number</u> can be counted or measured in appropriate units. The following kinds of numbers are real:

- <u>Zero</u> represents the absence of something. It is real because you can definitely measure when something is missing. For example, if you proceed to measure how much water is in a beaker, if the beaker is empty then the beaker clearly contains zero liters of water. As another example, if an ATM machine shows a balance of $0, it means that there is no money left in the account.
- <u>Whole numbers</u> include 1, 2, 3, 4, 5, 6, 7, 8, 9, 10, 11, 12, etc. Whole numbers are real because they can be counted. For example, if a bag contains apples, the number of apples can be determined simply by counting them.
- <u>Negative numbers</u> include $-1, -2, -3, -4, -5, -6, -7$, etc. Negative numbers are real because they can be counted backwards. For example, if you set up a coordinate system where $+y$ is upward and the origin lies on the ground, if a ball is dropped down a well that is 20 meters deep, the final position of the ball is $y = -20$ meters. As another example, if a person has no money and borrows $40, the person has a net worth of $-$$40; this person needs to earn $40 just to have no money at all. As yet another example, if a player's token lies on the 18th space of a board game and the player draws a card that says to go back 3 spaces, the token goes -3 spaces forward. Here, the negative sign indicates that the spaces will be subtracted: the token will be placed on the 15th space because $18 + (-3) = 18 - 3 = 15$.
- <u>Rational numbers</u> can be expressed as fractions of the form $\frac{a}{b}$, where a and b are both integers. Rational numbers include fractions like $\frac{4}{7}, \frac{8}{12}$ (which can be reduced to $\frac{2}{3}$ by dividing 8 and 12 each by 4), $\frac{3}{2}$ (which is called an improper fraction because the numerator is larger than the denominator), or $2\frac{1}{4}$ (which is a mixed fraction; it is two and a fourth). Whole numbers are rational because they equate to fractions with a denominator of one. For example, the number

5 can be expressed in the form $\frac{a}{b}$ with $a = 5$ and $b = 1$. Note that $\frac{5}{1} = 5$. Decimals like 0.25 and percents like 25% are also rational because they can be expressed as fractions. Note that $\frac{1}{4} = 0.25 = 25\%$. Even repeating decimals like 1.333333... where the 3 repeats forever (it can be expressed using an overbar as $1.\bar{3}$) are rational because they can be expressed as fractions where the numerator and denominator are both integers. Note that $\frac{4}{3} = 1.333333...$ with the 3 repeating forever. (For an example of a decimal that isn't rational, see the next bullet point.) Rational numbers are real because they can be measured. For example, if a pizza is cut into 8 slices and there are 3 slices remaining after dinner is eaten, then we can see that $\frac{3}{8}$ of the pizza remains. As another example, if salt is poured into a measuring cup until it reaches $\frac{2}{3}$ of a cup, we can measure that there are two-thirds of a cup of salt. As yet another example, if a stopwatch measures the time it takes for a ball to drop to be 0.425 seconds, this time is rational because the fraction $\frac{17}{40}$ is equal to the decimal 0.425.

- **Irrational numbers** can't be expressed in the form $\frac{a}{b}$ where both a and b are integers, yet they can be measured. When an irrational number is measured in decimal form, the digits go on forever without forming a repeating pattern. For example, $\sqrt{2}$ is an irrational number. In decimal form, $\sqrt{2}$ begins 1.41421356... and the digits continue endlessly without ever forming repeating pattern. (In contrast, the number 1.3636363636... where 36 repeats forever is a rational number equal to the fraction $\frac{15}{11}$.) It can be proven that $\sqrt{2}$ can't be expressed in the form $\frac{a}{b}$ where both a and b are integers.[1] Another example of an irrational number is the number π (represented by the lowercase Greek letter pi). If a circle has a diameter of one meter, the circle has a circumference equal to π meters. If the circumference is measured in decimal form, the number begins 3.14159265... with the digits continuing forever without any repeating pattern.

---

[1] If interested, search for Hippasus' proof that $\sqrt{2}$ can't be expressed as a ratio of two integers.

Complex Numbers Essentials Math Workbook with Answers

When the solution to an algebra problem is a **real number**, the problem can be modeled in the real world and the answer can be measured. The equations below provide examples from algebra for each kind of real number discussed previously.

- $x + 4 = 4$ has the real solution $x = 0$. This answer is zero. If an account has \$4 and you wish to add money to the account so that the balance is \$4, zero money needs to be added to the account.
- $x^2 = 9$ has the real solutions $x = 3$ and $x = -3$. These answers are integers. If a square has an area of 9 square meters, each side is 3 meters long.
- $x + 9 = 2$ has the real solution $x = -7$. This answer is negative. If you have a string that is 9 inches long and wish to add string to it so that the new length is 2 inches, you need to add -7 inches to the length. In this case, the minus sign indicates that 7 inches need to be subtracted from its length: $-7 + 9 = 2$.
- $8x = 6$ has the real solution $x = \frac{6}{8} = \frac{3}{4}$. This answer is rational. If you buy 8 oranges for \$6, then each orange costs \$$\frac{6}{8}$ = \$0.75 (which equals 75 cents).
- $x^2 = 3$ has the real solutions $x = \sqrt{3}$ and $x = -\sqrt{3}$. These answers are irrational. If a square field has an area of 3 square miles, then each side measures $\sqrt{3}$ miles. This number begins 1.73205080... with the digits continuing forever without forming a repeating pattern.

Now we'll consider an example where the solution is **not** real. Compare the two equations below.

$$x^2 = 4 \quad , \quad x^2 = -4$$

- The first equation, $x^2 = 4$, has real solutions: $x = 2$ or $x = -2$. If you square 2, you get 4. If you square -2, you also get 4 since $(-2)(-2) = 4$.
- The second equation, $x^2 = -4$ doesn't have a real solution. Any real number squared is nonnegative. Any positive number squared is positive; any negative number squared is also positive; zero squared is zero. There doesn't exist a real number whose square is negative. In an introductory algebra course, you would be taught that $x^2 = -4$ has **no solution**. However, it turns out that an entire subject of **complex numbers**, which have a variety of applications (see Chapter 17), has been developed around solutions to equations like $x^2 = -4$.

Chapter 1 – Imaginary Numbers

Suppose that there exists a number that solves the equation $x^2 = -1$. We know that the solution isn't real, so it makes sense to call the number **imaginary**. The number i is referred to as an **imaginary number** and solves the equation $x^2 = -1$. That is, when the imaginary number i is squared, we get minus one: $i^2 = -1$. Put another way, the square root of negative one is an imaginary number: $i = \sqrt{-1}$ (but see the note later in this chapter entitled "Be careful with the square roots of negative numbers").

Although i itself can't be directly measured, the imaginary number i turns out to be very useful in both math and physics. We'll work out one example later in this chapter that shows the value of complex numbers. For additional examples, see Chapter 17.

If k is a real number, then the solution to the equation $x^2 = -k^2$ is **imaginary**. The answers $x = ki$ and $x = -ki$ are both solutions to $x^2 = -k^2$. When we square both sides of $x = ki$, we get $x^2 = (ki)^2 = k^2 i^2 = -k^2$ because $i^2 = -1$. Similarly, when we square both sides of $x = -ki$, we get $x^2 = (-ki)^2 = (-1)^2 k^2 i^2 = (1)k^2(-1) = -k^2$. Note that if k is a real number, then ki is imaginary.

Let's see what happens when i is raised to a power:
- $i^0 = 1$. Recall from algebra that any nonzero number raised to the power of zero equals one: $x^0 = 1$ if $x \neq 0$.[2]
- $i^1 = i$. This follows from the rule that $x^1 = x$.
- $i^2 = -1$. This was our definition of the imaginary number. Put another way, i times i equals negative one.
- $i^3 = -i$. Multiply by i on both sides of $i^2 = -1$.
- $i^4 = 1$. Multiply by i on both sides of $i^3 = -i$. Note: $i(-i) = -i^2 = -(-1) = 1$. Alternatively, square both sides of $i^2 = -1$ to get $i^4 = 1$.

It will be really helpful to know that $i^0 = 1$, $i^1 = i$, $i^2 = -1$, $i^3 = -i$, and $i^4 = 1$ as well as you know your multiplication facts. These arise frequently with complex numbers.

---

[2] What if $x = 0$? In that case, 0^0 is indeterminate. Why? Recall the rule from algebra that $\frac{x^m}{x^n} = x^{m-n}$. Now let $m = n$ to see that $\frac{x^n}{x^n} = x^{n-n} = x^0$. When $x = 0$, we get $\frac{0^n}{0^n} = 0^0$. Since $\frac{0^n}{0^n} = \frac{0}{0}$ and since $\frac{0}{0}$ is indeterminate, that's why 0^0 is indeterminate; division by zero always poses a problem.

Any other power of i (which is a positive integer) can be related to $i^1 = i$, $i^2 = -1$, $i^3 = -i$, and $i^4 = 1$ as follows. Consider i^n, where $n > 0$ is an integer. Determine the **remainder** when n is divided by 4. For example, if $n = 27$, the remainder of $27 \div 4$ equals 3. (Why? The largest multiple of 4 that doesn't exceed 27 is $6 \times 4 = 24$. Subtract 24 from 27 to get a remainder of 3.) The remainder tells you the answer according to the cases below.

- If the remainder is zero, $i^n = 1$. For example, $i^{12} = 1$ because $12 \div 4$ doesn't have a remainder (12 is a multiple of 4, since $4 \times 3 = 12$).
- If the remainder is one, $i^n = i$. For example, $i^9 = i$ because the remainder of $9 \div 4$ equals one. (The largest multiple of 4 that doesn't exceed 9 is $2 \times 4 = 8$. Subtract 8 from 9 to get a remainder of 1.)
- If the remainder is two, $i^n = -1$. For example, $i^{14} = -1$ because the remainder of $14 \div 4$ equals two. (The largest multiple of 4 that doesn't exceed 14 is $3 \times 4 = 12$. Subtract 12 from 14 to get a remainder of 2.)
- If the remainder is three, $i^n = -i$. For example, $i^{23} = -i$ because the remainder of $23 \div 4$ equals three. (The largest multiple of 4 that doesn't exceed 23 is $5 \times 4 = 20$. Subtract 20 from 23 to get a remainder of 3.)

Why does dividing the exponent by 4 and finding the remainder work? It's because $i^4 = 1$. If you multiply by i^4, i^8, i^{12}, i^{16}, i^{20}, or any other power of i that is a multiple of 4, you are just multiplying by one. For example, $i^{20} = (i^4)^5 = 1^5 = 1$ according to the rule from algebra that $(x^m)^n = x^{mn}$. In the above bullet points, we are basically factoring out i^4's. For example, $i^{23} = i^{20+3} = i^{20} i^3 = (1) i^3 = -i$ since $i^{20} = (i^4)^5 = 1$.

Why use imaginary numbers?

We will work out one example in trigonometry now and illustrate more applications in Chapter 17. Suppose that you wish to derive trig identities for $\cos(x + y)$ and for $\sin(x + y)$. Take a moment to think about how you would do this without using any knowledge of complex numbers. Ordinarily, this is a done by drawing right triangles.[3]

---

[3] For example, see Trig Identities Practice Workbook with Answers by Chris McMullen.

It turns out that this same problem is much easier to solve using complex numbers. We will use Euler's formula (which we'll learn in Chapter 11), $e^{i\theta} = \cos\theta + i\sin\theta$. In terms of x or y, Euler's formula is $e^{ix} = \cos x + i\sin x$ or $e^{iy} = \cos y + i\sin y$. Multiply these two equations together:[4]

$$e^{ix}e^{iy} = (\cos x + i\sin x)(\cos y + i\sin y)$$

Recall the rule $t^a t^b = t^{a+b}$ from algebra and expand the right-hand side.

$$e^{ix+iy} = \cos x \cos y + i\cos x \sin y + i\sin x \cos y + i^2 \sin x \sin y$$

Recall that $i^2 = -1$. Also, factor out the i in the exponential.

$$e^{i(x+y)} = \cos x \cos y + i\cos x \sin y + i\sin x \cos y - \sin x \sin y$$

Now let $\theta = x + y$ in Euler's formula to get an equation for $e^{i\theta} = e^{i(x+y)}$:

$$e^{i(x+y)} = \cos(x+y) + i\sin(x+y)$$

Set the last two equations equal to one another.

$$\cos(x+y) + i\sin(x+y) = \cos x \cos y + i\cos x \sin y + i\sin x \cos y - \sin x \sin y$$

The above equation actually has two equations in one. For the equation to be true, the real parts of both sides must be equal and the imaginary parts of both sides must also be equal. This allows us to separate the above equation into two different equations.

$$\cos(x+y) = \cos x \cos y - \sin x \sin y$$
$$\sin(x+y) = \cos x \sin y + \sin x \cos y$$

Just like that, we have identities for $\cos(x+y)$ and $\sin(x+y)$. This example will probably make more sense when you've finished the book. (We'll see more problems like this in Chapter 17.) For now, the point is that there can be real value to working with imaginary numbers. Just because there doesn't exist a 'real' solution to the equation $x^2 = -1$ doesn't mean that the solution ($x = \pm i$) doesn't have value and usefulness. The entire subject of **quantum mechanics** is riddled with complex numbers, which may be the ultimate example of a significant application of complex numbers. (We'll briefly discuss a couple of examples of quantum mechanics in Chapter 17.)

---

[4] Can you multiply equations together? Yes. The product of the left-hand sides equals the product of the right-hand sides. If $a = b$ and $c = d$, it is certainly true that $ac = bd$. If needed, try it with numbers. Let $a = 2$ and $c = 3$. Then $ac = bd$ says that $(2)(3) = (2)(3)$.

Be careful with the square roots of negative numbers: For example, recall the rule $\sqrt{a^2} = \sqrt{a}\sqrt{a} = a$ from algebra, where a is a nonnegative number. If you apply this rule when a is negative, you get an interesting result. For example, try letting a equal minus one. In this case, you get $\sqrt{(-1)^2} = \sqrt{-1}\sqrt{-1} = -1$, while on the other hand, since $(-1)^2 = (-1)(-1) = 1$, you could alternative write this as $\sqrt{(-1)^2} = \sqrt{1} = 1$. One way to resolve this problem is to consider that $x = -1$ and $x = 1$ both solve the equation $x^2 = 1$. If you square root both sides of $x^2 = 1$, you get $\sqrt{x^2} = \pm 1$. There is an inherent plus or minus sign with square roots. In algebra, we tend to follow the convention that $\sqrt{y}$ implies the positive root only; if you want both roots, you would instead write $\pm\sqrt{y}$. (In contrast, when solving an equation of the form $y^2 = 3$, we would find both roots, writing the solution as $y = \pm\sqrt{3}$. Both roots solve the equation $y^2 = 3$.) Again in calculus, we tend to use $\sqrt{y}$ to imply the positive root only because calculus deals with functions, which are single-valued; you can't have a function with plus or minus signs. However, if you square root negative numbers, the plus or minus nature of square roots can be quite important. If you take only the positive root, you can easily 'prove' that 1 equals minus 1, which is sheer nonsense. For example, start with $i^2 = -1$. Make the substitution $i = \sqrt{-1}$ to get $\sqrt{-1}\sqrt{-1} = -1$. Use the rule that $\sqrt{a}\sqrt{a} = \sqrt{a^2}$ to get $\sqrt{(-1)^2} = -1$. Since $(-1)^2 = (-1)(-1) = 1$, it would seemingly follow that 1 equals -1. However, 1 obviously doesn't equal -1. Considering the plus or minus nature of square roots, what we have really proven is that $\pm 1 = \pm 1$. This 'proof' simply exploits the usual convention of taking positive roots in such a way that a missing alternative negative root makes a huge difference. Such things can happen when taking square roots of negative numbers. It's important to keep these ideas in mind when working with imaginary numbers.

One tip is to consider how i is really defined. We started with $i^2 = -1$. If you square root both sides of this equation, what you really get is $i = \pm\sqrt{-1}$. The second of these possible solutions was omitted in the above 'proof.' We will prefer to work with i than to write $\sqrt{-1}$. If a problem starts with $\sqrt{-1}$, beware that this really equals $\pm i$ (and not simply i). If you replace $\sqrt{-1}$ with i (rather than $\pm i$), you could run into trouble.

Chapter 1 – Imaginary Numbers

Example 1. Evaluate i^{28}.

Solution: The remainder of $28 \div 4$ is 0 (since $4 \times 7 = 28$). Since 28 is a multiple of 4, $i^{28} = 1$. Put another way, $i^{28} = (i^4)^7 = (1)^7 = 1$. Here, we used $i^4 = 1$ and the rule from algebra that $(x^m)^n = x^{mn}$.

Alternate solution: Raise both sides of $i^2 = -1$ to the power of 14 to get $(i^2)^{14} = (-1)^{14}$, which simplifies to $i^{28} = 1$. Recall from algebra that any even power of -1 equals one: $(-1)^2 = (-1)^4 = (-1)^6 = (-1)^8 = \cdots = 1$.

Example 2. Evaluate $\frac{1}{i}$.

Solution: Multiply the numerator and denominator each by i to get $\frac{1}{i}\frac{i}{i} = \frac{i}{i^2} = \frac{i}{-1} = -i$. Recall that $i^2 = -1$. Recall from algebra that $\frac{x}{-1} = -x$.

Example 3. Evaluate i^{39}.

Solution: The remainder of $39 \div 4$ is 3 (since $4 \times 9 = 36$ is the largest multiple of 4 that doesn't exceed 39 and since $39 - 36 = 3$). Thus, $i^{39} = i^{36+3} = i^{36}i^3 = (1)(-i) = -i$. We're basically saying that since $i^{36} = (i^4)^9 = 1$ and $i^3 = -i$, it follows that $i^{39} = -i$.

Alternate solution: Divide i^{40} by i. Since 40 is a multiple of 4, it follows from Example 1 that $i^{40} = 1$. Thus, $i^{39} = \frac{i^{40}}{i} = \frac{1}{i} = \frac{1}{i}\frac{i}{i} = \frac{i}{i^2} = \frac{i}{-1} = -i$ (see Example 2).

Example 4. Evaluate i^{62}.

Solution: The remainder of $62 \div 4$ is 2 (since $4 \times 15 = 60$ is the largest multiple of 4 that doesn't exceed 62 and since $62 - 60 = 2$). Thus, $i^{62} = i^{60+2} = i^{60}i^2 = (1)(-1) = -1$. We're basically saying that since $i^{60} = (i^4)^{15} = 1$ and $i^2 = -1$, it follows that $i^{62} = -1$.

Example 5. Evaluate $(-2i)^5$.

Solution: Recall the rule from algebra that $(ax)^n = a^n x^n$. Thus, $(-2i)^5 = (-2)^5 i^5 = (-1)^5 (2)^5 i^5 = (-1)(32)i^5 = -32i^5$. The remainder of $5 \div 4$ is 1 (since $4 \times 1 = 4$ is the largest multiple of 4 that doesn't exceed 5 and since $5 - 4 = 1$). Thus, $i^5 = i^1 = i$. Put another way, $i^5 = i^{4+1} = i^4 i^1 = (1)(i) = i$. Thus, $-32i^5 = -32i$. Recall from algebra that odd powers of -1 are $(-1)^3 = (-1)^5 = (-1)^7 = (-1)^9 = \cdots = -1$.

Chapter 1 Problems

Directions: Evaluate each expression.

(1) i^{19}

(2) i^{56}

(3) i^{33}

(4) i^{2002}

(5) $(-3i)^3$

(6) $(4i)^4$

(7) $\dfrac{1}{i^3}$

(8) $\left(\dfrac{2}{i}\right)^6$

(9) $\left(-\dfrac{5}{i}\right)^4$

2 Complex Numbers

A **complex number** consists of both real and imaginary parts. For example, $5 - 3i$ is a complex number; the real part is 5 and the imaginary part is -3. In Chapter 5, we will learn that the complex number $5 - 3i$ can be visualized as a point in the complex plane as follows: first go 5 units along the real axis and then go -3 units along the imaginary axis. Note that the number $5 - 3i$ cannot be simplified further, much as the number $3 + \sqrt{2}$ cannot be simplified further. Following are examples of complex numbers:

- $8 + 4i$
- $1 - i$
- $\frac{\sqrt{3}}{2} + \frac{i}{2}$
- $3i^2 + 5i$, which simplifies to $-3 + 5i$ since $i^2 = -1$ (recall Chapter 1)

As counterexamples, the examples below **don't** have **both** real **and** imaginary parts; each example below is either purely real or purely imaginary.

- $\frac{\sqrt{2}}{2}$ is purely real; it doesn't have an imaginary part.
- $(5 - \sqrt{3})i$ is purely imaginary; it doesn't have a real part. If we distribute, we get $5i - i\sqrt{3}$; every term has an i.
- $i^3 - 2i$ is purely imaginary. Since $i^3 = -i$ (recall Chapter 1), this simplifies to $-i - 2i = -3i$, which doesn't have a real part.

To determine whether a number is complex, purely real, or purely imaginary, follow these steps:

- If there are any powers of i (like i^3), simplify them like we did in Chapter 1.
- If needed, simplify every term until the term is either real (like 4, $-\frac{3}{5}$, or $\sqrt{2}$) or is proportional to i (like $-5i$, $\frac{i}{2}$, or $i\sqrt{5}$).
- If there are only real terms, the number is purely real.
- If there are no real terms (that is, every term is proportional to i), the number is purely imaginary.
- If there are both real terms and terms proportional to i, the number is complex.

Complex Numbers Essentials Math Workbook with Answers

Example 1. Is the expression below complex, purely real, or purely imaginary?
$$i^3 - 3i^2 + 5i$$
Solution: Recall from Chapter 1 that $i^3 = -i$ and $i^2 = -1$.
$$i^3 - 3i^2 + 5i = -i - 3(-1) + 5i = -i + 3 + 5i = 3 + 4i$$
In the last step, we combined the like terms $-i$ and $5i$. The answer is complex because the first term (3) is real while the second term ($4i$) is imaginary.

Example 2. Is the expression below complex, purely real, or purely imaginary?
$$\frac{1}{i} - i$$
Solution: Multiply the numerator and denominator of the first term each by i.
$$\frac{1}{i} - i = \frac{1}{i}\frac{i}{i} - i = \frac{i}{i^2} - i = \frac{i}{-1} - i = -i - i = -2i$$
Recall that $i^2 = -1$. Note that $\frac{i}{-1} = -\frac{i}{1} = -i$. The answer is purely imaginary because there isn't a real term. The only term ($-2i$) is proportional to i.

Example 3. Is the expression below complex, purely real, or purely imaginary?
$$5i^4 + i^2\sqrt{2}$$
Solution: Recall from Chapter 1 that $i^4 = 1$ and $i^2 = -1$.
$$5i^4 + i^2\sqrt{2} = 5(1) + (-1)\sqrt{2} = 5 - \sqrt{2}$$
The answer is purely real because $5 - \sqrt{2}$ doesn't include any imaginary parts. No term is proportional to i.

Example 4. Is the expression below complex, purely real, or purely?
$$\frac{i^{10}}{2} + \frac{i^7}{3}$$
Solution: First use the method that we learned in Chapter 1 to determine that $i^{10} = i^8 i^2 = (1)(-1) = -1$ and $i^7 = i^4 i^3 = (1)(-i) = -i$. (The remainder of $10 \div 4$ is 2 since $8 \div 4 = 2$ and $10 - 8 = 2$. The remainder of $7 \div 4$ is 3 since $7 - 4 = 3$.)
$$\frac{i^{10}}{2} + \frac{i^7}{3} = -\frac{1}{2} + \frac{-i}{3} = -\frac{1}{2} - \frac{1}{3}i$$
The answer is complex because the first term ($-1/2$) is real while the second term ($-i/3$) is imaginary.

Chapter 2 Problems

Directions: Simplify each expression as much as possible. Determine whether each expression is complex, purely real, or purely imaginary.

(1) $2i^2 + 3i^3 + 5i^5 + 8i^8 + 13i^{13}$

(2) $7i^{15} + 5i^{27} - 3i^{49} - i^{75}$

(3) $(2i)^8 - (3i)^6 + (4i)^4 - (6i)^2$

(4) $\frac{i}{8} + \left(\frac{i}{4}\right)^2 + \left(\frac{i}{2}\right)^3$

(5) $\frac{1}{2} - \frac{2}{i}$

(6) $\frac{5}{i^5} + \frac{3}{i^3} - \frac{1}{i}$

(7) $i^3\sqrt{3} + 2i^4 + 3i^5 + 4i^6\sqrt{6}$

(8) $9i^9 + i^7\sqrt{7} - 5i^5$

3 Complex Conjugates

From what we learned about complex numbers in Chapter 2, we can see that any **complex number** can be expressed in the form $x + iy$, where x and y are both nonzero real numbers. If a complex number has the form $x + iy$, its **complex conjugate** has the form $x - iy$, where the sign of the imaginary part is reversed. For example, for the complex number $2 + 5i$, its complex conjugate is $2 - 5i$.

There are two popular conventions for indicating a complex conjugate. Some texts use an **asterisk** (*), as in $(2 + 5i)^* = 2 - 5i$. Other texts use an **overbar**, as in $\overline{2 + 5i} = 2 - 5i$. Both methods are fairly common. In this text, we will use overbars. Let $z = x + iy$, where x and y are both real numbers.[1] Then the $\bar{z} = x - iy$ is the complex conjugate of z. (It's mutual; z is also the complex conjugate of $\bar{z}$. That is, z and $\bar{z}$ are each complex conjugates of one another.)

The main idea here is that if you want to find the complex conjugate of a number, all you need to do is reverse the sign of the imaginary part, as in these examples:
- The complex conjugate of $3 + 8i$ is equal to $3 - 8i$.
- The complex conjugate of $9 - 5i$ is equal to $9 + 5i$.
- The complex conjugate of $i - 1$ is equal to $-i - 1$.
- The complex conjugate of $-\frac{1}{2} - i\frac{\sqrt{3}}{2}$ is equal to $-\frac{1}{2} + i\frac{\sqrt{3}}{2}$.

A number and its complex conjugate share an important relationship. Here are a few of their fundamental properties:
- $z + \bar{z}$ is always real: $x + iy + (x - iy) = 2x$.
- $z - \bar{z}$ is always imaginary: $z - \bar{z} = x + iy - (x - iy) = 2iy$.
- $z\bar{z}$ is always real: $(x + iy)(x - iy) = x^2 - ixy + ixy - i^2y = x^2 + y^2$.

---

[1] In the first paragraph, we said "nonzero real numbers," but now it says "real numbers." What's the deal? If x and y are both **nonzero** real numbers, then $x + iy$ is **complex** (which is a word we used in the first paragraph). What we're saying in the current paragraph applies even if x or y happens to be zero. In these special cases, if y is zero, then $x + iy$ is purely real (so the number and its complex conjugate are both simply x), whereas if x is zero and y isn't, then $x + iy$ is purely imaginary.

Chapter 3 – Complex Conjugates

The last of these points, that $z\bar{z}$ (and also $\bar{z}z$) is always a real number, is particularly significant in both math and physics. For example, in quantum mechanics, the symbol ψ (which is the Greek letter psi) represents the wave function of a particle and can be found by solving Schrödinger's equation. In general, ψ is a complex number, but this isn't a problem because ψ itself can't be directly measured. What can be measured is $\psi\bar{\psi}$ (which is the product of the wave function and its complex conjugate), as this product turns out to be related to probabilities. Since $\psi\bar{\psi}$ is always a real number, even though ψ is often a complex number, the probabilities in quantum mechanics always turn out to be real numbers (and therefore measurable). We'll briefly consider the role of complex numbers in quantum mechanics in Chapter 17.

Tip: It's worth remembering that $(x + iy)(x - iy) = x^2 + y^2$ (as shown previously in the last bullet point). When multiplying a number by its complex conjugate, all you need to do is add the sum of the squares.

Example 1. Find the complex conjugate of $8 + 6i$.
Solution: Simply reverse the sign of the imaginary part: $\overline{8 + 6i} = 8 - 6i$.

Example 2. Find the complex conjugate of $12 - 3i$.
Solution: Simply reverse the sign of the imaginary part: $\overline{12 - 3i} = 12 + 3i$.

Example 3. Find the complex conjugate of $\frac{2}{i}$.
Solution: First express the number in the form $x + iy$ by multiplying and numerator and denominator each by i to get $\frac{2}{i}\frac{i}{i} = \frac{2i}{i^2} = -2i$. Now reverse the sign of the imaginary part: $\overline{-2i} = 2i$. (Since $-2i$ is purely imaginary, there isn't a real part.)

Example 4. Evaluate $(4 + 2i)\overline{4 + 2i}$.
Solution: Use the formula $(x + iy)(x - iy) = x^2 + y^2$.
$$(4 + 2i)\overline{4 + 2i} = (4 + 2i)(4 - 2i) = 4^2 + 2^2 = 16 + 4 = 20$$

Example 5. Evaluate $(3 - i)\overline{3 - i}$.
Solution: Use the formula $(x + iy)(x - iy) = x^2 + y^2$.
$$(3 - i)\overline{3 - i} = (3 - i)(3 + i) = 3^2 + (-1)^2 = 9 + 1 = 10$$

Chapter 3 Problems

Directions: Find each complex conjugate.

(1) $\overline{4 + 2i}$

(2) $\overline{7 - 5i}$

(3) $\overline{i + 1}$

(4) $\overline{-11 + 6i}$

(5) $\overline{8i + 15}$

(6) $\overline{(5i)^3}$

(7) $\overline{-3 - 9i}$

(8) $\overline{3 + \frac{6}{i}}$

(9) $\overline{i^{22} + 2i^{17} - 3i^8}$

Directions: Evaluate each expression.

(10) $(6 - 3i)\overline{6 - 3i}$

(11) $(i + 2)\overline{i + 2}$

(12) $i\,\overline{i}$

(13) $(5 - 12i)\overline{5 - 12i}$

(14) $(-7 + 8i)\overline{-7 + 8i}$

(15) $(-9 - 4i)\overline{-9 - 4i}$

(16) $(i^{12} - 2i^9 + 3i^6)\overline{i^{12} - 2i^9 + 3i^6}$

(17) $(2 + \frac{1}{i})\overline{2 + \frac{1}{i}}$

4 Complex Arithmetic

In this chapter, we will learn how to add, subtract, multiply, or divide complex numbers.

To add or subtract complex numbers, the main idea is to **combine like terms**. We use this same strategy in algebra to add $5x + 4$ and $3x + 2$, for example. In this case, $5x$ and $3x$ are like terms because each term has the same variable raised to the first power, and 4 and 2 are like terms because each is a constant.
$$5x + 4 + 3x + 2 = (5x + 3x) + (4 + 2) = 8x + 6$$
We similarly use this same strategy to add $7 + 4\sqrt{2}$ and $5 + 3\sqrt{2}$. In this case, 7 and 5 are like terms because each is rational, and $4\sqrt{2}$ and $3\sqrt{2}$ are like terms because each is proportional to the same irrational number ($\sqrt{2}$).
$$7 + 4\sqrt{2} + 5 + 3\sqrt{2} = 12 + 7\sqrt{2}$$
When we **combine like terms**, what we really do is **factor**. When we combine $5x$ and $3x$ to make $8x$, we're really just factoring out an x like this: $5x + 3x = (5 + 3)x = 8x$. Similarly, when we combine $4\sqrt{2}$ and $3\sqrt{2}$, we're factoring out $\sqrt{2}$ like this: $4\sqrt{2} + 3\sqrt{2} = (4 + 3)\sqrt{2} = 7\sqrt{2}$.

The same idea of combining like terms (or, equivalently, factoring) applies when we **add or subtract complex numbers**. For example, to add $3 + 6i$ to $8 + 9i$, the 3 and 8 are like terms because they are both real, while the $6i$ and $9i$ are like terms because they are both imaginary.
$$3 + 6i + 8 + 9i = (3 + 8) + (6i + 9i) = 11 + 15i$$
When we add $6i$ and $9i$, we basically factor out an i like this: $6i + 9i = (6 + 9)i = 15i$. If you subtract two complex numbers, it works much the same way, but be sure to **distribute** the minus sign. For example, consider $7 + 4i$ minus $2 + 3i$:
$$7 + 4i - (2 + 3i) = 7 + 4i - 2 - 3i = (7 - 2) + (4i - 3i) = 5 + i$$
(Of course, $1i = i$ just like $1x = x$ in algebra; a coefficient of 1 is implied.) When there is a negative sign in parentheses and the parentheses are being subtracted, the two minus signs make a plus sign, just like in algebra. For example, consider $6 + 7i$ minus $4 - 2i$. Watch how subtracting $-2i$ makes a positive $2i$:

$$6 + 7i - (4 - 2i) = 6 + 7i - 4 - (-2i) = 6 + 7i - 4 + 2i$$
$$= (6 - 4) + (7i + 2i) = 2 + 9i$$

The main idea here is that $7i - (-2i) = 7i + 2i = 9i$. Two minus signs effectively make a plus sign.

To multiply two complex numbers, distribute like you normally would in algebra and then combine like terms like we did when we added complex numbers. First recall how to multiply two algebraic expressions like $3x + 2$ and $5x + 4$. First, **distribute** the $3x$ to both the $5x$ and the 4, and then distribute the 2 to the $5x$ and the 4:

$$(3x + 2)(5x + 4) = 3x(5x) + 3x(4) + 2(5x) + 2(4)$$
$$= 15x^2 + 12x + 10x + 8 = 15x^2 + 22x + 8$$

The $12x$ and $10x$ are like terms, so we combined them by factoring out the x. That is, $12x + 10x = (12 + 10)x = 22x$.

The same idea applies when **multiplying complex numbers**. As an example, we will multiply $3 + 6i$ by $4 + 3i$. First distribute the 3 to both the 4 and the $3i$, and then also distribute the $6i$ to the 4 and the $3i$:

$$(3 + 6i)(4 + 3i) = 3(4) + 3(3i) + 6i(4) + 6i(3i) = 12 + 9i + 24i + 18i^2$$
$$= 12 + 33i + 18(-1) = -6 + 33i$$

Recall that $i^2 = -1$. We factored out an i with $9i + 24i = (9 + 24)i = 33i$.

To **divide two complex numbers**, follow these steps. We will illustrate these steps for the example of dividing $2 + 3i$ by $4 - 2i$.

- Express the division problem as a fraction. The numerator is divided by the denominator. We will write the division problem $(2 + 3i) \div (4 - 2i)$ as:
$$\frac{2 + 3i}{4 - 2i}$$
- Multiply the numerator and denominator each by the **complex conjugate** (recall Chapter 3) of the denominator. Why? The new denominator will be real.
$$\frac{2 + 3i}{4 - 2i}\left(\frac{4 + 2i}{4 + 2i}\right)$$
- Multiply the complex numbers in the numerator. Also, use the formula in the denominator for multiplying a number by its complex conjugate (Chapter 3).

Chapter 4 – Complex Arithmetic

$$\frac{2(4) + 2(2i) + 3i(4) + 3i(2i)}{4^2 + 2^2} = \frac{8 + 4i + 12i + 6i^2}{16 + 4} = \frac{8 + 16i - 6}{20} = \frac{2 + 16i}{20}$$

- If the denominator and every term of the numerator share a common factor, factor it out to reduce the fraction. For example, the denominator (20) and each term of the numerator (2 and $16i$) are evenly divisible by 2, so we will factor a 2 out of the numerator and denominator. The 2's will cancel out.

$$\frac{2 + 16i}{20} = \frac{2(1 + 8i)}{2(10)} = \frac{1 + 8i}{10} = \frac{1}{10} + \frac{8i}{10} = \frac{1}{10} + \frac{4}{5}i$$

Example 1. Evaluate $3 + 4i + 2 + 5i$.

Solution: Combine like terms:

$$3 + 4i + 2 + 5i = (3 + 2) + (4i + 5i) = 5 + 9i$$

We basically factored out an i since $4i + 5i = (4 + 5)i = 9i$.

Example 2. Evaluate $8 + 6i - (5 + 2i)$.

Solution: Distribute the minus sign and combine like terms.

$$8 + 6i - (5 + 2i) = 8 + 6i - 5 - 2i = (8 - 5) + (6i - 2i) = 3 + 4i$$

Example 3. Evaluate $5 - 7i - (9 - 8i)$.

Solution: Distribute the minus sign and combine like terms. Subtracting a negative sign equates to addition (two minus signs make a plus sign).

$$5 - 7i - (9 - 8i) = 5 - 7i - 9 - (-8i) = 5 - 7i - 9 + 8i$$
$$= (5 - 9) + (-7i + 8i) = -4 + i$$

Notes: $-7 + 8 = 1$. Just as $1x = x$ in algebra, here $1i = i$.

Example 4. Evaluate $(6 + 2i)(3 - 5i)$.

Solution: Distribute the 6 to both the 3 and $-5i$, and also distribute the $2i$ to both the 3 and $-5i$. This makes four terms. Then use $i^2 = -1$ and combine like terms.

$$(6 + 2i)(3 - 5i) = 6(3) + 6(-5i) + 2i(3) + 2i(-5i)$$
$$= 18 - 30i + 6i - 10i^2 = 18 - 24i - 10(-1) = 18 - 24i + 10 = 28 - 24i$$

Note: The minus 10 and minus one combine together to make plus 10.

Example 5. Evaluate $(6 - 4i)^2$.

Solution: Multiply $(6 - 4i)$ by itself using the method from Example 4.
$$(6 - 4i)(6 - 4i) = 6(6) + 6(-4i) - 4i(6) - 4i(-4i)$$
$$= 36 - 24i - 24i + 16i^2 = 36 - 48i + 16(-1) = 36 - 48i - 16 = 20 - 48i$$

Example 6. Evaluate $(9 + 6i) \div (3 - 2i)$.

Solution: First write this as a fraction with $9 + 6i$ over $3 - 2i$.
$$\frac{9 + 6i}{3 - 2i}$$
The complex conjugate (Chapter 3) of the denominator is $3 + 2i$. Multiply by $3 + 2i$ in both the numerator and denominator. (We may do this because $\frac{3+2i}{3+2i} = 1$.)
$$\frac{9 + 6i}{3 - 2i}\left(\frac{3 + 2i}{3 + 2i}\right)$$
Multiply the expressions in the numerator like Example 4. In the denominator, use the formula $(x + iy)(x - iy) = x^2 + y^2$ from Chapter 3.
$$\frac{9(3) + 9(2i) + 6i(3) + 6i(2i)}{3^2 + 2^2} = \frac{27 + 18i + 18i + 12i^2}{9 + 4} = \frac{27 + 36i - 12}{13}$$
$$= \frac{15 + 36i}{13} = \frac{15}{13} + \frac{36}{13}i$$
Since the denominator (13) doesn't share a common factor with the terms from the numerator, this answer can't be reduced. Contrast this with the next example.

Example 7. Evaluate the expression below.
$$\frac{15}{3 + i}$$
Solution: The complex conjugate (Chapter 3) of the denominator is $3 - i$. Multiply by $3 - i$ in both the numerator and denominator.
$$\frac{15}{3 + i}\left(\frac{3 - i}{3 - i}\right) = \frac{15(3) + 15(-i)}{3^2 + 1^2} = \frac{45 - 15i}{9 + 1} = \frac{45 - 15i}{10}$$
The denominator (10) and each term of the numerator (45 and also $-15i$) is evenly divisible by 5. Therefore, we may reduce this fraction by factoring a 5 out of both the numerator and denominator. The 5 cancels out:
$$\frac{45 - 15i}{10} = \frac{5(9 - 3i)}{5(2)} = \frac{9 - 3i}{2} = \frac{9}{2} - \frac{3}{2}i$$

Chapter 4 Problems

Directions: Perform the indicated arithmetic operation.

(1) $10 + 6i + 12 - 4i$

(2) $(2 + 7i) - (5 + 3i)$

(3) $(-4 + 9i) + (-6 + 8i)$

(4) $11 + 3i - (6 - 9i)$

(5) $(15 - 10i) - (7 + 6i)$

(6) $24 - 18i + (16 - 12i)$

(7) $21 - 12i - (21 + 12i)$

(8) $(17 - 11i) - (-8 + 4i)$

(9) $(32 + 48i) + (-17 + 24i)$

(10) $(-60 + 40i) - (-20 + 80i)$

(11) $(8 + 4i)(5 + 6i)$

(12) $(9 + 5i)(4 - 7i)$

(13) $(6 - 7i)(8 - 3i)$

(14) $(7 - 5i)(5 + 9i)$

(15) $(3 - 9i)\overline{8 - 4i}$

(16) $(8 + 5i)^2$

(17) $(2 - i)^3$

(18) $(2 - i)^4$

Chapter 4 – Complex Arithmetic

(19) $(4 - 8i) \div (2 + 5i)$

(20) $(15 - 25i) \div (2 - i)$

(21) $\dfrac{6}{1+i}$

(22) $\dfrac{5+10i}{4-3i}$

(23) $\dfrac{-120+240i}{-2+4i}$

(24) $\dfrac{(8+i)(4-i)}{(3-2i)^2}$

5 The Complex Plane

The **complex plane** provides a way to geometrically visualize complex numbers:
- The horizontal axis (x) represents the real numbers. A purely real number lies on the x-axis.
- The vertical axis (y) represents the imaginary numbers. A purely imaginary number lies on the y-axis.
- As usual, the point of intersection is called the **origin**. This is the point (0,0).
- A number that doesn't lie on either the x- or the y-axis is complex. It has both real and imaginary parts.
- The coordinates of the complex number $z = x + iy$ are (x, y).

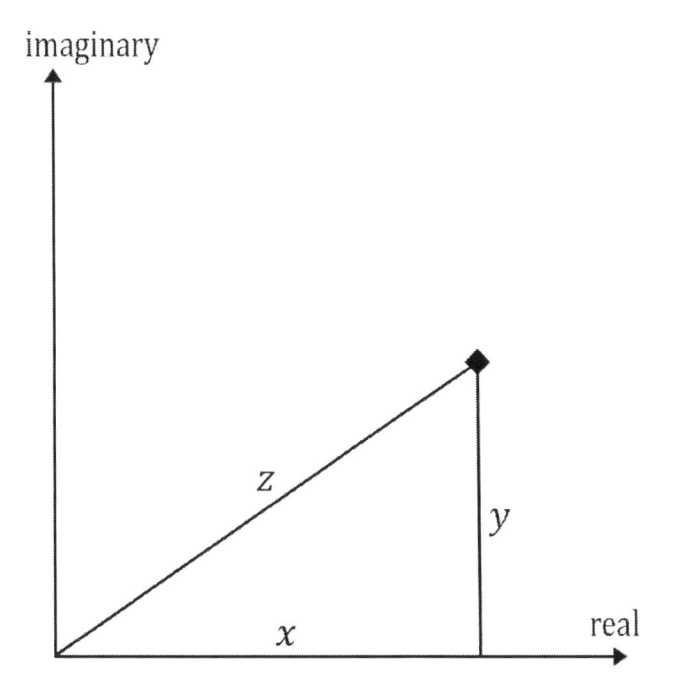

The graph above is referred to as an **Argand diagram**. The diamond (♦) marks the location of the point (x, y), which represents the complex number $z = x + iy$. You should notice that graphing a point in the complex plane really is no different than graphing a point (x, y) in the usual xy plane. The only distinction is the interpretation that x is the real part and y is the imaginary part of a complex number.

If you know about **vectors**, z is basically a vector with components of x and y. (In AC circuits, we call z a **phasor**, which is basically a vector where the vertical component is the imaginary part of a complex number.) If you don't know about vectors, don't feel left out. All we need to know about them in this book is the magnitude (which we will call the modulus in the context of complex numbers), direction, and how they relate to the components, and we'll learn these things in the next couple of chapters. In this chapter, we will focus on understanding how the real and imaginary parts of a complex number relate to the complex plane. This should seem easy because it isn't any different form plotting (x, y) in the usual xy plane. See the examples below.

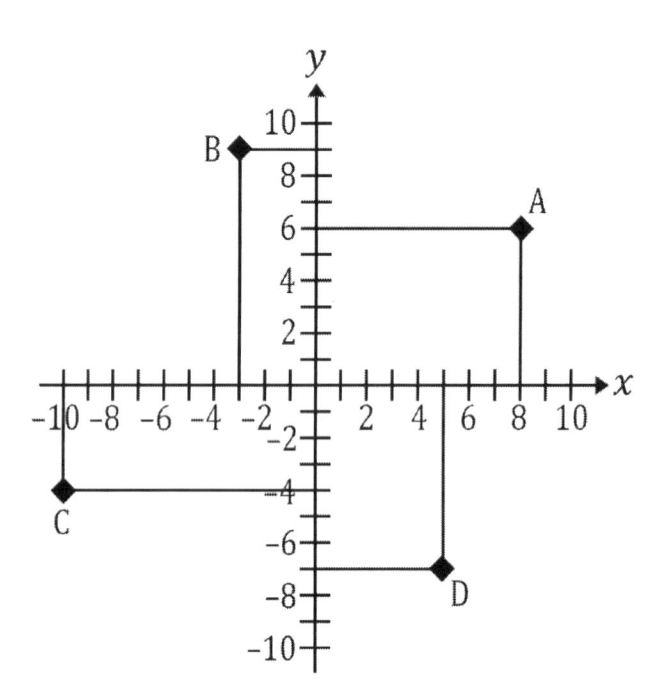

Example 1. Four different points are plotted in the complex plane above. Write each point as a complex number in the form $x + iy$.

Solution: Right/left is real; up/down is imaginary. (Left or down are negative.)
- A: 8 units right, 6 units up: $8 + 6i$ (the first number is left/right)
- B: 3 units left (this is minus), 9 units up: $-3 + 9i$
- C: 10 units left (this is minus), 4 units down (also minus): $-10 - 4i$
- D: 5 units right, 7 units down (this is minus): $5 - 7i$

Example 2. Given the complex number $z = 7 + 4i$, plot $z, \bar{z}, -z,$ and $-\bar{z}$. How are these complex numbers related in the complex plane?

Solution:
- $z = 7 + 4i$ is 7 units right and 4 units up (the first number is right/left)
- $\bar{z} = 7 - 4i$ is 7 units right and 4 units down
  $\bar{z}$ is the complex conjugate to z (recall Chapter 4)
  $\bar{z}$ is the reflection of z about the real axis
- $-z = -7 - 4i$ is 7 units left and 4 units down
  if you rotate z through 180° (about the origin), you get $-z$
  if you connect z and $-z$ by a straight line (see the diagonal below), this line will pass through the origin
  $-z$ is the reflection of $\bar{z}$ about the imaginary axis
- $-\bar{z} = -7 + 4i$ is 7 units left and 4 units up
  if you rotate $\bar{z}$ through 180° (about the origin), you get $-\bar{z}$
  $-\bar{z}$ is the reflection of z about the imaginary axis
  $-\bar{z}$ is the reflection of $-z$ about the real axis

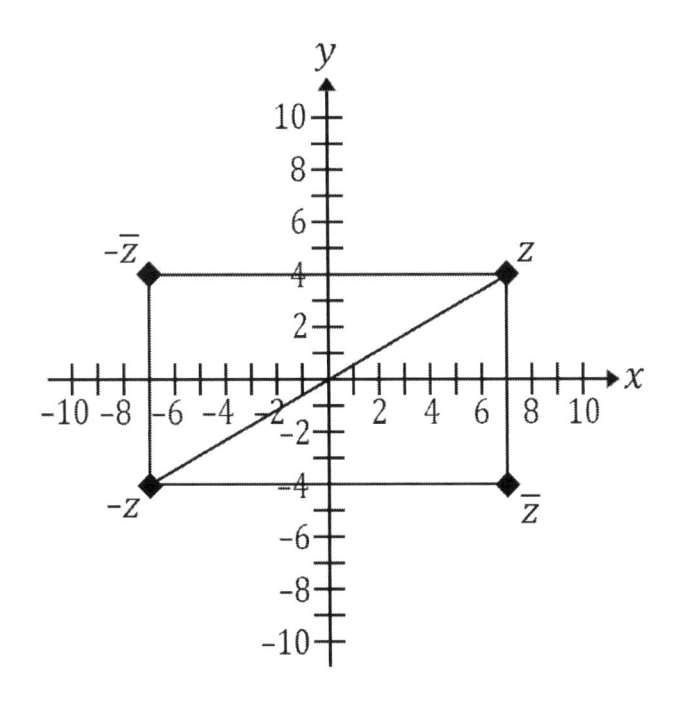

Chapter 5 Problems

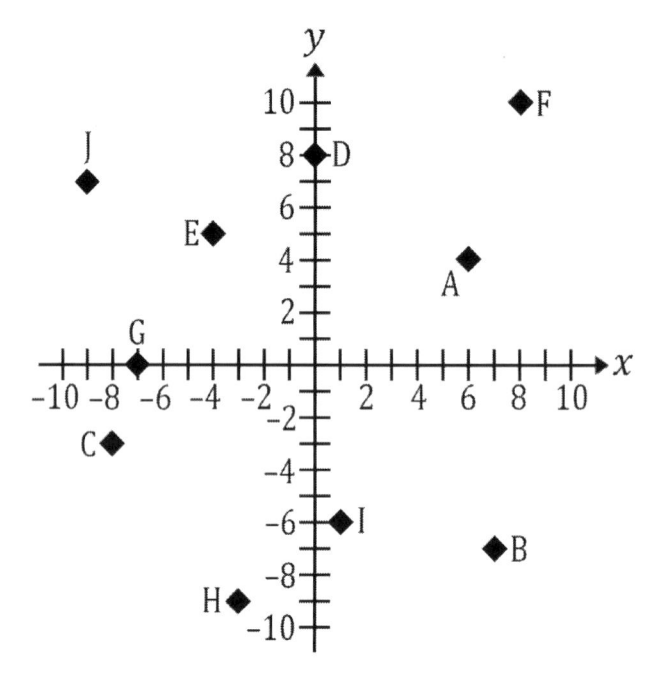

(1) Ten points are plotted in the complex plane above. Write each point as a complex number in the form $x + iy$.

(2) For the complex number $z = -8 + 3i$, plot z, $\bar{z}$, $-z$, and $-\bar{z}$.

6 The Modulus

The Argand diagram below shows the complex number $z = x + iy$, corresponding to the point (x, y) marked by a diamond (♦). The distance from the origin to z is called the **modulus** and is denoted by absolute value symbols: $|z|$. The modulus[1] can be found using the Pythagorean theorem for the right triangle below: $|z| = \sqrt{x^2 + y^2}$.

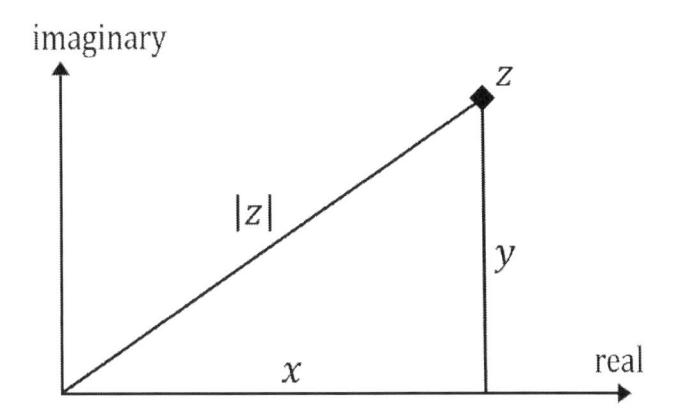

The term **modulus** refers to $|z| = \sqrt{x^2 + y^2}$, which is a distance (it is the hypotenuse of the triangle), whereas the term **modulus-square** refers to $|z|^2 = x^2 + y^2$, which is the square of the modulus.

If you remember the formulas from the previous chapters, you should recognize the formula for the modulus-square. It's the same as the product of a complex number with its complex conjugate (Chapter 3): $z\bar{z} = (x + iy)(x - iy) = x^2 + y^2$. Therefore, another way to write the formula for the modulus square is $|z|^2 = z\bar{z}$. Similarly, the formula for the modulus can be expressed as $|z| = \sqrt{z\bar{z}}$. Since x and y are each real, the quantity $z\bar{z} = x^2 + y^2$ is always real and nonnegative, which should make sense because $|z|$ is the distance from the origin to (x, y).

---

[1] In the context of vectors, we would refer to the modulus as the **magnitude** of z. A vector is a quantity that has both a magnitude and a direction. In complex numbers, the magnitude (or modulus) represents how far the point is from the origin, while the direction indicates which way the point lies from the origin. We'll focus on the modulus in this chapter, and discuss the direction in Chapter 7.

Chapter 6 – The Modulus

Example 1. Find the modulus of $6 - 8i$.
Solution: Use the formula for the modulus. Note that $x = 6$ and $y = -8$.
$$|z| = \sqrt{x^2 + y^2} = \sqrt{6^2 + (-8)^2} = \sqrt{36 + 64} = \sqrt{100} = 10$$

Example 2. Find the modulus of $1 + i$.
Solution: Use the formula for the modulus. Compare $1 + i$ with $z = x + iy$ to see that $x = 1$ and $y = 1$. (It would be a big mistake to think that y equals i. Remember that x and y are real, whereas $x + iy$ is complex.)
$$|z| = \sqrt{x^2 + y^2} = \sqrt{1^2 + 1^2} = \sqrt{1 + 1} = \sqrt{2}$$

Example 3. Find the modulus of $-3 + i\sqrt{3}$.
Solution: Use the formula for the modulus. Note that $x = -3$ and $y = \sqrt{3}$.
$$|z| = \sqrt{x^2 + y^2} = \sqrt{(-3)^2 + \left(\sqrt{3}\right)^2} = \sqrt{9 + 3} = \sqrt{12} = \sqrt{4}\sqrt{3} = 2\sqrt{3}$$

Recall the rule from algebra that $\left(\sqrt{a}\right)^2 = \sqrt{a}\sqrt{a} = a$. In the last steps, we factored the **perfect square** 4 out of 12 to write $\sqrt{12}$ as $2\sqrt{3}$; we used the rule $\sqrt{ab} = \sqrt{a}\sqrt{b}$. Many instructors prefer the form $2\sqrt{3}$ instead of $\sqrt{12}$; when it is possible to factor out a perfect square, they prefer for their students to do so.

Example 4. Given $a = 3 + 5i$ and $b = 7 + 2i$, find (A) $|a + b|$ and (B) $|a - b|$.
Solution: First find $a + b$ and $a - b$ using the method from Chapter 4, and then find the modulus.
(A) First find
$$a + b = 3 + 5i + 7 + 2i = (3 + 7) + (5i + 2i) = 10 + 7i$$
Use $x = 10$ and $y = 7$.
$$|a + b| = \sqrt{x^2 + y^2} = \sqrt{10^2 + 7^2} = \sqrt{100 + 49} = \sqrt{149}$$
(B) First find
$$a - b = 3 + 5i - (7 + 2i) = 3 + 5i - 7 - 2i = (3 - 7) + (5i - 2i) = -4 + 3i$$
Use $x = -4$ and $y = 3$.
$$|a + b| = \sqrt{x^2 + y^2} = \sqrt{(-4)^2 + 3^2} = \sqrt{16 + 9} = \sqrt{25} = 5$$

Chapter 6 Problems

Directions: Given $a = 12 + 5i$ and $b = 8 - 15i$, evaluate each expression below.

(1) $|a|$

(2) $|b|$

(3) $|a + b|$

(4) $|a - b|$

(5) $|3a - 2b|$

(6) $|ab|$

(7) $\left|\dfrac{a}{b}\right|$

(8) $|a|^2$

(9) $|a^2|$

Chapter 6 – The Modulus

Directions: Given $t = 2 - i$, $u = 2 + i\sqrt{3}$, and $w = \frac{2}{i}$, evaluate each expression below.

(10) $|t|$

(11) $|u|$

(12) $|w|$

(13) $|t + \overline{w}|$

(14) $|u - w\sqrt{3}|$

(15) $|\overline{t} + \overline{u}\sqrt{3} + \overline{w}|$

(16) $|t\overline{u}|$

(17) $\dfrac{|u|^2}{u}$

(18) $\dfrac{|u|^2}{t}$

7 Polar Form

Any complex number z may be expressed in the **Cartesian** (or **rectangular**) form $z = x + iy$, where x and y are real numbers. The Cartesian form separates the complex number into real and imaginary parts. An alternative way to express a complex number is to use polar form. The **polar** form also consists of two parts: the **modulus** $|z|$ and the **argument** θ. The **modulus** (Chapter 6) is the magnitude of the complex number, which indicates how far the complex number lies from the origin in the complex plane (Chapter 5). The **argument** θ is the direction of the complex number, which is an angle indicating how far counterclockwise the complex number is from the $+x$-axis (which is the same way that angles are measured in trigonometry). See the diagram below.

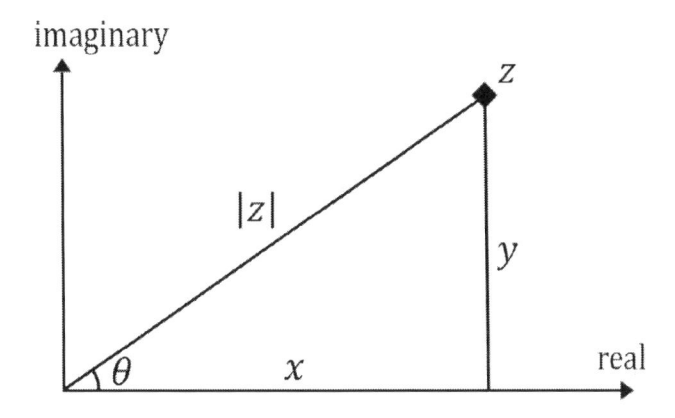

The relationship between the modulus and argument to the components x and y is the same as the relationship between polar coordinates and Cartesian coordinates. (If you've never seen polar coordinates before, don't worry. The equations that we will need appear below.) Applying trig to the right triangle above,[1] we get the following equations which relate $|z|$ and θ to x and y:

$$x = |z| \cos \theta$$
$$y = |z| \sin \theta$$

---

[1] Trig gives you $\cos \theta = \frac{x}{|z|}$ and $\sin \theta = \frac{y}{|z|}$. If you multiply by $|z|$ on both sides of these equations, you get $x = |z| \cos \theta$ and $y = |z| \sin \theta$.

Chapter 7 – Polar Form

The previous pair of equations is helpful if you know the modulus $|z|$ and argument θ of the complex number and wish to find x and y. If instead you know x and y and wish to find $|z|$ and θ, use the equations below. The first equation is the equation for the modulus from Chapter 6. The second equation comes from the definition of the tangent[2] of an angle as the opposite side over the adjacent side.

$$|z| = \sqrt{x^2 + y^2}$$
$$\theta = \tan^{-1}\left(\frac{y}{x}\right)$$

When using the second equation above, look at the signs of x and y to put θ in the correct quadrant. Actually, if you use a calculator, there is a simple trick to get θ right. If x is positive, the calculator correctly places θ in Quadrant I or IV, but if x is negative, you need to add 180° (or π radians) to what the calculator gives you to put the angle in Quadrant II or III. (This 180° trick works for **inverse tangent**; it's not as simple if you take an inverse sine or cosine.) So all you need to do is look at the sign of x. If x is negative, add 180° to the calculator's answer; if $x > 0$, do nothing. See the examples.

If you add or subtract a multiple of 360° to θ, you get an equivalent answer for θ (since trig functions are periodic). For example, −60° and 300° are equivalent arguments.

Here is another important calculator tip: Check if your calculator's mode is in **degrees** or **radians**. (Recall that π radians equates to 180°. To convert from radians to degrees, multiply by $\frac{180}{\pi}$.)

Note the following special cases:
- For a real number, $y = 0$ such that $|z| = |x|$. A real number lies on the x-axis (which is the real axis). If $x > 0$, then $\theta = 0°$, but if $x < 0$, then $\theta = 180°$.
- For a purely imaginary number, $x = 0$ such that $|z| = |y|$. A purely imaginary number lies on the y-axis (which is the imaginary axis). If $y > 0$, then $\theta = 90°$, but if $y < 0$, then $\theta = 270°$ (which is equivalent to −90°).

---

[2] The tangent gives you $\tan \theta = \frac{y}{x}$. To solve for θ, take the inverse tangent (also referred to as the arctangent) of both sides to get $\theta = \tan^{-1}\left(\frac{y}{x}\right)$. This uses the identity $\tan^{-1} \tan(\theta) = \theta$.

36

Complex Numbers Essentials Math Workbook with Answers

Example 1. Find the modulus and argument for $1 - i\sqrt{3}$.

Solution: Compare $1 - i\sqrt{3}$ with $z = x + iy$ to identify that $x = 1$ and $y = -\sqrt{3}$. Use the formulas that have $|z|$ and θ isolated.

$$|z| = \sqrt{x^2 + y^2} = \sqrt{1^2 + (-\sqrt{3})^2} = \sqrt{1+3} = \sqrt{4} = 2$$

$$\theta = \tan^{-1}\left(\frac{y}{x}\right) = \tan^{-1}\left(\frac{-\sqrt{3}}{1}\right) = -60°$$

Since $x > 0$ and $y < 0$, the answer for θ lies in Quadrant IV. You may optionally add $360°$ to the angle to express it as $300°$. The angles $-60°$ and $300°$ are equivalent.

Example 2. Find the modulus and argument for $-3 + 4i$.

Solution: Compare $-3 + 4i$ with $z = x + iy$ to identify that $x = -3$ and $y = 4$. Use the formulas that have $|z|$ and θ isolated.

$$|z| = \sqrt{x^2 + y^2} = \sqrt{(-3)^2 + 4^2} = \sqrt{9 + 16} = \sqrt{25} = 5$$

$$\theta = \tan^{-1}\left(\frac{y}{x}\right) = \tan^{-1}\left(\frac{4}{-3}\right) = -53.13° + 180° = 126.87°$$

Since $x < 0$ and $y > 0$, the answer for θ lies in Quadrant II. A standard calculator only gives inverse tangents in Quadrants I or IV. Since $x < 0$, you must add $180°$ to the calculator's answer in order to place the angle in the correct quadrant. Make sure that your calculator is in **degrees** mode if you want an answer in degrees.

Example 3. A complex number has a modulus of 6 and an argument of $135°$. Express the complex number in Cartesian form.

Solution: We are given $|z| = 6$ and $\theta = 135°$. Use the formulas that have x and y isolated. Make sure that your calculator is in degrees mode. (Students who are fluent in trig can do this problem without a calculator, since the reference angle for $135°$ equals $45°$.)

$$x = |z|\cos\theta = 6\cos 135° = 6\left(-\frac{\sqrt{2}}{2}\right) = -3\sqrt{2}$$

$$y = |z|\sin\theta = 6\sin 135° = 6\left(\frac{\sqrt{2}}{2}\right) = 3\sqrt{2}$$

The Cartesian form of the complex number is $-3\sqrt{2} + 3i\sqrt{2}$.

Chapter 7 Problems

Directions: Find the modulus and argument for each number.

(1) $10 + 24i$

(2) $7 - 7i$

(3) $-\sqrt{3} + i$

(4) $5i$

(5) $-\sqrt{2} - i\sqrt{6}$

(6) $(4i)^4$

(7) $2 - i$

(8) -11

(9) $\frac{9}{i}$

Directions: Express each number in Cartesian form.

(10) modulus $= 8$
argument $= 60°$

(11) modulus $= \sqrt{3}$
argument $= 330°$

(12) modulus $= 13$
argument $= 90°$

(13) modulus $= \sqrt{2}$
argument $= 225°$

(14) modulus $= 17$
argument $= 128°$

(15) modulus $= 4.8$
argument $= 196°$

(16) modulus $= 25$
argument $= 304°$

(17) modulus $= 15$
argument $= 180°$

(18) modulus $= \sqrt{6}$
argument $= 240°$

8 Real and Imaginary Parts

It is sometimes helpful to find just the real or imaginary part of a complex number. The notation Re(z) means just the real part of z, while Im(z) means just the imaginary part of z. If you put a complex number in the Cartesian form $z = x + iy$, where x and y are both real, then Re(z) = x and Im(z) = y. For example, for the complex number $7 - 2i$, the real and imaginary parts are Re($7 - 2i$) = 7 and Im($7 - 2i$) = -2.

Example 1. Find the real and imaginary parts of $6 - i\sqrt{3}$.

Solution: Compare $6 - i\sqrt{3}$ with $z = x + iy$ (where x and y are both real) to identify that Re(z) = x = 6 and Im(z) = y = $-i\sqrt{3}$.

Example 2. Find the real and imaginary parts of $\frac{2}{1+i}$.

Solution: First carry out the division according to Chapter 4. Multiply by the complex conjugate of the denominator.

$$\frac{2}{1+i}\left(\frac{1-i}{1-i}\right) = \frac{2(1-i)}{1^2 + 1^2} = \frac{2(1-i)}{2} = 1 - i$$

Now find the real and imaginary parts: Re($1 - i$) = 1 and Im($1 - i$) = -1.

Example 3. Given $u = 2 - 3i$ and $w = 5 + i$, find $\frac{\text{Re}(uw)}{\text{Re}(u)+\text{Im}(w)}$.

Solution: First find the indicated product:

$$uw = (2 - 3i)(5 + i) = 2(5) + 2(i) - 3i(5) - 3i(i) = 10 + 2i - 15i - 3i^2$$
$$uw = 10 - 13i - 3(-1) = 10 - 13i + 3 = 13 - 13i$$

Now find the real and imaginary parts.

$$\frac{\text{Re}(uw)}{\text{Re}(u) + \text{Im}(w)} = \frac{\text{Re}(13 - 13i)}{\text{Re}(2 - 3i) + \text{Im}(5 + i)} = \frac{13}{2 + 1} = \frac{13}{3}$$

Chapter 8 Problems

Directions: Given $a = 4 - 3i$, $b = 12 + 5i$, and $c = \frac{2}{i}$, evaluate each expression below.

(1) $\text{Re}(a)\text{Re}(b)$

(2) $\text{Re}(ab)$

(3) $5\text{Re}(b) - 10\text{Im}(a)$

(4) $\text{Im}(c)$

(5) $\text{Im}(a + b + c)$

(6) $\dfrac{\text{Re}(a)\text{Im}(b)}{\text{Re}(b)+\text{Im}(a)}$

(7) $\text{Im}\left(\dfrac{b}{a}\right) - \dfrac{\text{Im}(b)}{\text{Im}(a)}$

(8) $\dfrac{b+\overline{b}}{2} - \text{Re}(b)$

(9) $\dfrac{b-\overline{b}}{2i} + \text{Im}(b)$

9 Quadratic Roots

Recall from algebra that a quadratic equation in the standard form $az^2 + bz + c$ is solved by the **quadratic formula**:
$$z = \frac{-b \pm \sqrt{b^2 - 4ac}}{2a}$$
The **discriminant** refers to the part under the radical: $b^2 - 4ac$. In general, there are two solutions to the quadratic equation, one for the + sign and the other for the minus sign (of the $\pm$ sign in the numerator). These two solutions are called **roots**. If the constants a, b, and c are **real**, then the discriminant determines whether the roots to the quadratic equation are real, imaginary, or complex, as follows.

- If $b^2 - 4ac > 0$ (meaning that $b^2 > 4ac$), then both roots are real.
- If $b^2 - 4ac = 0$ (meaning that $b^2 = 4ac$), then both roots are the same and they are real. Since they are the same, this is called a **double root**.
- If $b^2 - 4ac < 0$ (meaning that $b^2 < 4ac$), then neither root is real. If $b = 0$, both roots are purely imaginary. If $b \neq 0$, both roots are complex. In either case, the two roots are complex conjugates of one another.
- However, if any of the constants a, b, and c **isn't real**, then the above points don't apply. In this case, you need to examine the entire quadratic formula (not just the discriminant) to determine whether or not the roots are real. (In this case, if there are two complex roots, they may not be complex conjugates. It's also possible to have one complex root and one real root in this case.)

The quadratic formula solves every case of the quadratic equation. For some values of the constants a, b, and c, it is easier to **factor** the quadratic equation than it is to use the quadratic formula; for other cases, the quadratic formula is easier. Following are a few examples showing when it is reasonably easy to factor the quadratic. (Factoring is easier for some students than for others. It is a challenge for some students because you have to try a few different combinations to get it right. If the right combination doesn't come easily to you, maybe you just haven't had enough practice for it to click, but don't worry. You don't have to factor. You can use the quadratic formula instead.)

- $4z^2 - 3 = 0$ is easily solved since it doesn't have a middle term. Simply add 3 to both sides, divide by 4, and square root both sides: $z = \pm\sqrt{\frac{3}{4}} = \pm\frac{\sqrt{3}}{2}$. There are two real roots; one is positive, the other is negative.

- $2z^2 + 18 = 0$ can be solved the same way. Subtract 18 from both sides, divide by 2, and square root both sides: $z = \pm\sqrt{-\frac{18}{2}} = \pm\sqrt{-9} = \pm 3i$. There are two purely imaginary roots. Here, the discriminant is $b^2 - 4ac = 0^2 - 4(2)(18) = -144$. When the discriminant is negative (and the coefficients are real), the roots aren't real.

- $z^2 - z - 6 = 0$ can be factored as $(z-3)(z+2) = 0$, which has two real roots[1]: $z = 3$ and $z = -2$.

- $10z^2 + 4z - 32 = 0$ can be factored as $(2z+4)(5z-8)$, which has two real roots[2]: $2z + 4 = 0$ leads to $z = -2$ and $5z - 8 = 0$ leads to $z = \frac{8}{5}$.

- $z^2 + 2iz + 15 = 0$ can be factored as $(z - 3i)(z + 5i)$, which has two imaginary roots[3]: $z = 3i$ and $z = -5i$.

- $3z^2 + 4z + 5 = 0$ or $z^2 + z\sqrt{3} - 6$ aren't easy to factor. For cases that don't factor easily, it's convenient to use the quadratic formula.

---

[1] The 6 can be factored as (6)(1), (3)(2), (2)(3), or (1)(6). Since it is negative, one factor must be positive and the other factor must be negative, like (–3)(2) or like (1)(–6). The combination $(z - 3)(z + 2)$ works: $(z - 3)(z + 2) = z^2 + 2z - 3z - 6 = z^2 - z - 6$.

[2] The 10 can be factored as (10)(1), (5)(2), (2)(5), or (1)(10). The 32 has several factors, such as (1)(32), (4)(8), (8)(4), or (16)(2). Since the 10 is positive, it either has two positive or two negative factors. Since the 32 is negative, one of its factors must be negative and the other positive. The combination $(2z + 4)(5z - 8)$ works: $(2z + 4)(5z - 8) = 10z^2 - 16z + 20z - 32 = 10z^2 + 4z - 32$.

[3] The 15 can be factored as (15)(1), (5)(3), (3)(5), or (1)(15). The presence of the i suggests imaginary roots. Including the imaginary number, the positive 15 is made from factors with opposite signs, allowing for $i^2 = -1$. The combination $(z - 3i)(z + 5i)$ works: $(z - 3i)(z + 5i) = z^2 + 5iz - 3iz - 15i^2 = z^2 + 2iz + 15$. The rules regarding the discriminant don't apply because $b = 5i$ is a non-real constant.

Example 1. Find all of the roots of $3z^2 = 75$.

Solution: Since there is no linear term ($b = 0$), it is easy to solve this equation using algebra. Divide by 3 on both sides to get $z^2 = \frac{75}{3} = 25$. Square root both sides. Include a $\pm$ sign because $(-5)^2$ and 5^2 both equal 25. There are two real roots: $z = \pm 5$. Check the answer. Plug it into the given equation: $3z^2 = 3(\pm 5)^2 = 3(25) = 75$ ✓.

Example 2. Find all of the roots of $4z^2 + 8 = 0$.

Solution: Since there is no linear term ($b = 0$), it is easy to solve this equation using algebra. Subtract 8 from both sides to get $4z^2 = -8$ and divide by 4 on both sides to get $z^2 = -\frac{8}{4} = -2$. Square root both sides. Include a $\pm$ sign. Also, include an i because $i^2 = -1$. There are two purely imaginary roots: $z = \pm i\sqrt{2}$.

Check: $4z^2 + 8 = 4(\pm i\sqrt{2})^2 + 8 = 4(i^2)(\pm\sqrt{2})^2 + 8 = 4(-1)(2) + 8 = 0$ ✓.

Example 3. Find all of the roots of $z^2 + 4z = 12$.

Solution: First subtract 12 from both sides to put the equation in **standard form**. This gives $z^2 + 4z - 12 = 0$. This equation is simple enough to factor (but if you prefer not to, see the alternate solution): $(z + 6)(z - 2) = 0$. There are two real roots: $z = -6$ and $z = 2$. There are two real roots because the discriminant is positive:
$$b^2 - 4ac = 4^2 - 4(1)(-12) = 16 + 48 = 64$$
Alternate solution: Once the equation is in **standard form**, $z^2 + 4z - 12 = 0$, identify $a = 1$, $b = 4$, and $c = -12$. Use the quadratic formula.
$$z = \frac{-b \pm \sqrt{b^2 - 4ac}}{2a} = \frac{-4 \pm \sqrt{4^2 - 4(1)(-12)}}{2(1)} = \frac{-4 \pm \sqrt{16 + 48}}{2}$$
$$z = \frac{-4 \pm \sqrt{64}}{2} = \frac{-4 \pm 8}{2} = -2 \pm 4$$
$$z = -2 - 4 = -6 \quad \text{or} \quad z = -2 + 4 = 2$$
Check: $z^2 + 4z = (-6)^2 + 4(-6) = 36 - 24 = 12$ and $2^2 + 4(2) = 4 + 8 = 12$ ✓.

Chapter 9 – Quadratic Roots

Example 4. Find all of the roots of $9z^2 = 30z - 9$.

Solution: First subtract $30z$ and add 9 on both sides to put the equation in **standard form**. This gives $9z^2 - 30z + 9 = 0$. This equation factors as $(3z - 1)(3z - 9) = 0$. There are two[4] real roots: $3z - 1 = 0$ leads to $z = \frac{1}{3}$ and $3z - 9 = 0$ leads to $z = \frac{9}{3} = 3$. There are two real roots because the discriminant is positive:

$$b^2 - 4ac = (-30)^2 - 4(9)(9) = 900 - 324 = 576$$

Alternate solution: Once the equation is in **standard form**, $9z^2 - 30z + 9 = 0$, identify $a = 9$, $b = -30$, and $c = 9$. Use the quadratic formula.

$$z = \frac{-b \pm \sqrt{b^2 - 4ac}}{2a} = \frac{-(-30) \pm \sqrt{(-30)^2 - 4(9)(9)}}{2(9)} = \frac{30 \pm \sqrt{900 - 324}}{18}$$

$$z = \frac{30 \pm \sqrt{576}}{18} = \frac{30 \pm 24}{18}$$

$$z = \frac{30 - 24}{18} = \frac{6}{18} = \frac{1}{3} \quad \text{or} \quad z = \frac{30 + 24}{18} = \frac{54}{18} = 3$$

Check: $9z^2 = 9\left(\frac{1}{3}\right)^2 = \frac{9}{9} = 1$ agrees with $30z - 9 = 30\left(\frac{1}{3}\right) - 9 = 10 - 9 = 1$ and $9z^2 = 9(3)^2 = 9(9) = 81$ agrees with $30z - 9 = 30(3) - 9 = 90 - 9 = 81$ ✓.

Example 5. Find all of the roots of $2z^2 + 5iz + 3 = 0$.

Solution: This equation factors[5] as $(2z - i)(z + 3i) = 0$. There are two imaginary roots: $2z - i = 0$ leads to $z = \frac{i}{2}$ and $z + 3i = 0$ leads to $z = -3i$. (Since the coefficient of z, $5i$, is imaginary, the usual rules regarding the discriminant don't apply.)

Alternate solution: Identify $a = 2$, $b = 5i$, and $c = 3$. Use the quadratic formula.

$$z = \frac{-b \pm \sqrt{b^2 - 4ac}}{2a} = \frac{-5i \pm \sqrt{(5i)^2 - 4(2)(3)}}{2(2)} = \frac{-5i \pm \sqrt{-25 - 24}}{4}$$

$$z = \frac{-5i \pm \sqrt{-49}}{4} = \frac{-5i \pm \sqrt{49}\sqrt{-1}}{4} = \frac{-5i \pm 7i}{4}$$

---

[4] Since $(3z - 1)(3z - 9) = 0$, it follows that either $3z - 1 = 0$ or $3z - 9 = 0$. That is, the only way that the product of two numbers can be zero is if (at least) one of the numbers is zero.

[5] Here, $-i(3i) = -3i^2 = -3(-1) = 3$. With the imaginary number, the factors have opposite signs and use $i^2 = -1$ to make positive 3.

44

$$z = \frac{-5i - 7i}{4} = \frac{-12i}{4} = -3i \quad \text{or} \quad z = \frac{-5i + 7i}{4} = \frac{2i}{4} = \frac{i}{2}$$

Check: $2z^2 + 5iz + 3 = 2(-3i)^2 + 5i(-3i) + 3 = 2(-3)^2i^2 - 15i^2 + 3 = 2(9)(-1)$
$-15(-1) + 3 = -18 + 15 + 3 = 0$ and $2z^2 + 5iz + 3 = 2\left(\frac{i}{2}\right)^2 + 5i\left(\frac{i}{2}\right) + 3 =$
$2\left(\frac{1}{2}\right)^2 i^2 + \frac{5}{2}i^2 + 3 = 2\left(\frac{1}{4}\right)(-1) + \frac{5}{2}(-1) + 3 = -\frac{1}{2} - \frac{5}{2} + \frac{6}{2} = 0 \checkmark$.

Example 6. Find all of the roots of $z^2 + 5 = 4z$.

Solution: First subtract $4z$ from both sides to put the equation in **standard form**. This gives $z^2 - 4z + 5 = 0$. Now identify $a = 1$, $b = -4$, and $c = 5$. Use the quadratic formula.

$$z = \frac{-b \pm \sqrt{b^2 - 4ac}}{2a} = \frac{-(-4) \pm \sqrt{(-4)^2 - 4(1)(5)}}{2(1)} = \frac{4 \pm \sqrt{16 - 20}}{2}$$

$$z = \frac{4 \pm \sqrt{-4}}{2} = \frac{4 \pm \sqrt{4}\sqrt{-1}}{2} = \frac{4 \pm 2i}{2} = 2 \pm i$$

$$z = 2 - i \quad \text{or} \quad z = 2 + i$$

There are two complex roots, which are complex conjugates of one another because the discriminant is negative:

$$b^2 - 4ac = (-4)^2 - 4(1)(5) = 16 - 20 = -4$$

Check: $z^2 + 5 = (2 - i)^2 + 5 = (2 - i)(2 - i) + 5 = 2(2) + 2(-i) - i(2) - i(-i) + 5$
$= 4 - 2i - 2i + i^2 + 5 = 4 - 4i - 1 + 5 = 8 - 4i$ agrees with $4z = 4(2 - i) = 8 - 4i$
and $z^2 + 5 = (2 + i)^2 + 5 = (2 + i)(2 + i) + 5 = 2(2) + 2(i) + i(2) + i(i) + 5 =$
$4 + 2i + 2i + i^2 + 5 = 4 + 4i - 1 + 5 = 8 + 4i$ agrees with $4z = 4(2 + i) = 8 + 4i \checkmark$.

Chapter 9 Problems

Directions: Evaluate each expression.

(1) $5z^2 = 80$

(2) $3z^2 + 12 = 0$

(3) $6z^2 - 2 = 0$

(4) $8z^2 = -18$

(5) $z^2 + 48 = 16z$

(6) $6z^2 + z = 2$

(7) $10z^2 = 9z + 40$

(8) $4z^2 + 12z + 9 = 0$

(9) $z^2 + z\sqrt{2} = 4$

(10) $z^2 + 8iz = 15$

(11) $12z^2 + 23iz + 9 = 0$

(12) $z^2 + 36i = 9iz + 4z$

(13) $z^2 + 13 = 6z$

(14) $z^2 + 11 = 8z$

(15) $z^2 + 7 = 4z$

(16) $4z^2 + 34 = 20z$

10 De Moivre's Theorem

The relationship between the polar form and Cartesian form of complex numbers (Chapter 7) leads to two useful equations: de Moivre's theorem and Euler's formula. We will look at de Moivre's theorem now and discuss Euler's formula[1] in Chapter 11.

According to **de Moivre's theorem**,
$$(\cos\theta + i\sin\theta)^n = \cos(n\theta) + i\sin(n\theta)$$
The left hand-side involves $\cos\theta + i\sin\theta$. These are the Cartesian components of a complex number with a modulus of one ($|z| = 1$). Recall from Chapter 7 that if you are given a complex number with a modulus of $|z| = 1$ and an argument θ, you would find the Cartesian components as $x = |z|\cos\theta = (1)\cos\theta = \cos\theta$ and $y = |z|\sin\theta = (1)\sin\theta = \sin\theta$, such that the Cartesian form of the complex number would be $z = x + iy = \cos\theta + i\sin\theta$. Thus, the left-hand side is basically z^n for a complex number with a modulus of $|z| = 1$.

Why is de Moivre's theorem true? One way to see this is to work it out for the first few integer values of n and observe a pattern that develops.

- $n = 0$ is trivial: $(\cos\theta + i\sin\theta)^0 = \cos 0 + i\sin 0$ reduces to $1 = \cos 0$.
- $n = 1$ is also trivial: $(\cos\theta + i\sin\theta)^1 = \cos\theta + i\sin\theta$.
- For $n = 2$, the left-hand side is
$$(\cos\theta + i\sin\theta)^2 = \cos^2\theta + 2i\cos\theta\sin\theta + i^2\sin^2\theta$$
$$(\cos\theta + i\sin\theta)^2 = \cos^2\theta + 2i\sin\theta\cos\theta - \sin^2\theta$$
  Recall the trig identities $\cos(2\theta) = \cos^2\theta - \sin^2\theta$ and $\sin(2\theta) = 2\sin\theta\cos\theta$. These identities allow us to write the right-hand side as
$$\cos(2\theta) + i\sin(2\theta) = \cos^2\theta - \sin^2\theta + 2i\sin\theta\cos\theta$$
  This shows that de Moivre's theorem is true for $n = 2$.

---

[1] Textbooks where calculus is a prerequisite tend to derive Euler's formula first and then use Euler's formula to derive de Moivre's theorem. This has the advantage that it isn't necessary to use the method of 'induction' and that it can be proven for any value of n (not just integers). Since this book is designed to be accessible to students with a precalculus background, we did this in the opposite order. However, we will briefly see the calculus method (without doing the calculus) in Chapter 12 (see the Maclaurin series).

Chapter 10 – De Moivre's Theorem

- For $n = 3$, we'll begin with the right-hand side, $\cos(3\theta) + i\sin(3\theta)$. We'll use the angle sum identities from trig, $\cos(\alpha + \beta) = \cos\alpha\cos\beta - \sin\alpha\sin\beta$ and $\sin(\alpha + \beta) = \sin\alpha\cos\beta + \sin\beta\cos\alpha$, with $\alpha = 2\theta$ and $\beta = \theta$.
$$\cos(3\theta) + i\sin(3\theta) = \cos(2\theta + \theta) + i\sin(2\theta + \theta)$$
$$= \cos(2\theta)\cos\theta - \sin(2\theta)\sin\theta + i\sin(2\theta)\cos\theta + i\sin\theta\cos(2\theta)$$
Replace the minus sign with i^2.
$$\cos(3\theta) + i\sin(3\theta) = \cos(2\theta)\cos\theta + i^2\sin(2\theta)\sin\theta + i\sin(2\theta)\cos\theta + i\sin\theta\cos(2\theta)$$
Now factor out $\cos(2\theta)$ and $i\sin(2\theta)$. Note the i being factored with $\sin(2\theta)$.
$$\cos(3\theta) + i\sin(3\theta) = \cos(2\theta)(\cos\theta + i\sin\theta) + i\sin(2\theta)(\cos\theta + i\sin\theta)$$
Go a step further, factoring out $(\cos\theta + i\sin\theta)$.
$$\cos(3\theta) + i\sin(3\theta) = [\cos(2\theta) + i\sin(2\theta)](\cos\theta + i\sin\theta)$$
Using de Moivre's theorem for $n = 2$, we may write this as
$$\cos(3\theta) + i\sin(3\theta) = (\cos\theta + i\sin\theta)^2(\cos\theta + i\sin\theta)$$
Finally, we see that de Moivre's theorem is true for $n = 3$:
$$\cos(3\theta) + i\sin(3\theta) = (\cos\theta + i\sin\theta)^3$$

- The same steps that we just went through for $n = 3$ can be generalized to show that if de Moivre's theorem holds for n, it will also hold for $n + 1$. We start with
$$\cos[(n + 1)\theta] + i\sin[(n + 1)\theta] = \cos(n\theta + \theta) + i\sin(n\theta + \theta)$$
Use the angle sum identities from trig. (Recall the previous bullet point.)
$$\cos[(n + 1)\theta] + i\sin[(n + 1)\theta]$$
$$= \cos(n\theta)\cos\theta - \sin(n\theta)\sin\theta + i\sin(n\theta)\cos\theta + i\sin\theta\cos(n\theta)$$
$$= \cos(n\theta)\cos\theta + i^2\sin(n\theta)\sin\theta + i\sin(n\theta)\cos\theta + i\sin\theta\cos(n\theta)$$
We replaced the minus sign with i^2. Now factor out $\cos(n\theta)$ and $i\sin(n\theta)$.
$$\cos[(n + 1)\theta] + i\sin[(n + 1)\theta] = \cos(n\theta)(\cos\theta + i\sin\theta) + i\sin(n\theta)(\cos\theta + i\sin\theta)$$
Go a step further, factoring out $(\cos\theta + i\sin\theta)$.
$$\cos[(n + 1)\theta] + i\sin[(n + 1)\theta] = [\cos(n\theta) + i\sin(n\theta)](\cos\theta + i\sin\theta)$$
Rewrite this, using de Moivre's theorem for n.
$$\cos[(n + 1)\theta] + i\sin[(n + 1)\theta] = (\cos\theta + i\sin\theta)^n(\cos\theta + i\sin\theta)$$
Finally, we see that if de Moivre's theorem is true for n, it is also true for $n + 1$.
$$\cos[(n + 1)\theta] + i\sin[(n + 1)\theta] = (\cos\theta + i\sin\theta)^{n+1}$$
The above equation demonstrates de Moivre's theorem using the method of induction for the case where n is an integer (though the theorem holds in general).

In practice, it is often convenient to use another form of de Moivre's theorem, which we will consider now. If we multiply both sides of de Moivre's theorem by $|z|^n$, we get
$$(|z|\cos\theta + i|z|\sin\theta)^n = |z|^n \cos(n\theta) + i|z|^n \sin(n\theta)$$
In this form, the left-hand side is z^n for any complex number, regardless of what its modulus is. We will use this form of de Moivre's theorem in the following ways in the examples and exercises:

- Given a complex number in the Cartesian form $x + iy$, we can find $(x + iy)^n$. The way to do this is to find the modulus $|z|$ and argument θ like we did in Chapter 7. Then apply de Moivre's theorem. (For a large value of n, this is easier than multiplying $(x + iy)$ by itself repeatedly.[2])
$$(x + iy)^n = |z|^n \cos(n\theta) + i|z|^n \sin(n\theta)$$

- Given a complex number in the Cartesian form $x + iy$, we can find $\sqrt{x + iy}$. The way to do this is to set $n = 2$ in de Moivre's theorem and square root both sides. (If you want cube roots or other roots besides square roots, see Chapter 13.) We will call this modulus $|u|$ and this argument φ for this case (because we will get another modulus and argument later.)
$$(|u|\cos\varphi + i|u|\sin\varphi)^2 = |u|^2 \cos(2\varphi) + i|u|^2 \sin(2\varphi)$$
Now let $|u|^2 = |z|$ and let $\theta = 2\varphi$ (we will identify $|z|$ and θ as another modulus and argument).
$$\left(\sqrt{|z|}\cos\frac{\theta}{2} + i\sqrt{|z|}\sin\frac{\theta}{2}\right)^2 = |z|\cos\theta + i|z|\sin\theta$$
Square root both sides.
$$\sqrt{|z|}\cos\frac{\theta}{2} + i\sqrt{|z|}\sin\frac{\theta}{2} = \pm\sqrt{|z|\cos\theta + i|z|\sin\theta}$$
Given $x + iy$, find the modulus and argument. Then the square root of $x + iy$ is:
$$\sqrt{|z|}\cos\frac{\theta}{2} + i\sqrt{|z|}\sin\frac{\theta}{2} = \pm\sqrt{x + iy}$$
If $a = b$, it follows that $b = a$. Thus, we may rewrite the above formula as:
$$\pm\sqrt{x + iy} = \sqrt{|z|}\cos\frac{\theta}{2} + i\sqrt{|z|}\sin\frac{\theta}{2}$$
See Example 2.

---

[2] But if you know the binomial theorem from algebra, the multiplication isn't so bad.

Chapter 10 – De Moivre's Theorem

Example 1. Use trigonometry to evaluate $(\sqrt{3} - i)^5$.

Solution: Let $z = x + iy = \sqrt{3} - i$, such that $x = \sqrt{3}$ and $y = -1$. Find the modulus and argument like we did in Chapter 7.

$$|z| = \sqrt{x^2 + y^2} = \sqrt{(\sqrt{3})^2 + (-1)^2} = \sqrt{3 + 1} = \sqrt{4} = 2$$

$$\theta = \tan^{-1}\left(\frac{y}{x}\right) = \tan^{-1}\left(\frac{-1}{\sqrt{3}}\right) = -30°$$

Since $x > 0$ and $y < 0$, the answer for θ lies in Quadrant IV. (You may optionally add $360°$ to the angle to express it as $330°$. The angles $-30°$ and $330°$ are equivalent.) Now we may apply de Moivre's theorem with $n = 5$ as follows:

$$(|z|\cos\theta + i|z|\sin\theta)^5 = |z|^5 \cos(5\theta) + i|z|^5 \sin(5\theta)$$

Note that the left-hand side is simply $(x + iy)^5 = (\sqrt{3} - i)^5$, since $x = |z|\cos\theta = \sqrt{3}$ and $y = i|z|\sin\theta = -1$.

$$(\sqrt{3} - i)^5 = 2^5 \cos[5(-30°)] + 2^5 i \sin[5(-30°)] = 32\cos(-150°) + 32i\sin(-150°)$$

$$(\sqrt{3} - i)^5 = 32\left(-\frac{\sqrt{3}}{2}\right) + 32i\left(-\frac{1}{2}\right) = -16\sqrt{3} - 16i$$

Example 2. Evaluate $\sqrt{-65 - 72i}$. (You may recall that we encountered this problem in the alternate solution to Exercise 12 in Chapter 9. It will be simpler now.)

Solution: Let $z = x + iy = -65 - 72i$, such that $x = -65$ and $y = -72$. Find the modulus and argument like we did in Chapter 7.

$$|z| = \sqrt{x^2 + y^2} = \sqrt{(-65)^2 + (-72)^2} = \sqrt{4225 + 5184} = \sqrt{9409} = 97$$

$$\theta = \tan^{-1}\left(\frac{y}{x}\right) = \tan^{-1}\left(\frac{-72}{-65}\right) = \tan^{-1}\left(\frac{72}{65}\right) \approx 47.925° + 180° \approx 227.925°$$

Since $x < 0$ and $y < 0$, the answer for θ lies in Quadrant III. As discussed in Chapter 7, when $x < 0$ we must add $180°$ to the calculator's answer in order to put the angle in the right quadrant. Now we may apply de Moivre's theorem with $n = 2$ as follows:

$$\pm\sqrt{x + iy} = \sqrt{|z|}\cos\frac{\theta}{2} + i\sqrt{|z|}\sin\frac{\theta}{2}$$

The left-hand side is simply $\pm\sqrt{-65 - 72i}$. (The reason for the $\pm$ is that there are two answers for a square root. We'll address this below.)

$$\pm\sqrt{-65-72i} \approx \sqrt{97}\cos\frac{227.925°}{2} + i\sqrt{97}\sin\frac{227.925°}{2}$$

$$\pm\sqrt{-65-72i} \approx 9.84886\cos 113.963° + 9.84886i \sin 113.963°$$

$$\pm\sqrt{-65-72i} \approx 9.84886(-0.40615) + 9.84886i(0.91381)$$

$$\pm\sqrt{-65-72i} = -4 + 9i$$

$$\sqrt{-65-72i} = \pm(-4 + 9i)$$

$$\sqrt{-65-72i} = -4 + 9i \quad \text{or} \quad \sqrt{-65-72i} = 4 - 9i$$

The $\pm$ gave us two possible answers. Why? Consider the problem $t^2 = 25$, which has two answers because $(-5)^2 = 25$ and $5^2 = 25$.

Check the answer: If we multiply the answer by itself, it should equal $-65 - 72i$.

$$(-4+9i)(-4+9i) = -4(-4) - 4(9i) + 9i(-4) + 9i(9i)$$

$$(-4+9i)(-4+9i) = 16 - 36i - 36i + 81i^2 = 16 - 72i - 81 = -65 - 72i \checkmark$$

$$(4-9i)(4-9i) = 4(4) + 4(-9i) - 9i(4) - 9i(-9i)$$

$$(4-9i)(4-9i) = 16 - 36i - 36i + 81i^2 = 16 - 72i - 81 = -65 - 72i \checkmark$$

Chapter 10 Problems

Directions: Use trigonometry to evaluate each expression.

(1) $(3-3i)^4$

(2) $(i-2)^6$

(3) $(\sqrt{2}+i\sqrt{6})^5$

(4) $(-3-4i)^7$

(5) $(i - \sqrt{2})^8$

(6) $\left(\sqrt{3} - 1 - \frac{i}{2}\right)^3$

(7) $\sqrt{5 + 12i}$

(8) $\sqrt{i}$

(9) $\sqrt{-i}$

(10) $\sqrt{15-8i}$

(11) $\sqrt{28i-45}$

(12) $\sqrt{1+i\sqrt{3}}$

11 Euler's Formula

According to **Euler's formula**, $e^{i\theta} = \cos\theta + i\sin\theta$, where $e^{i\theta}$ is referred to as an exponential and where $e \approx 2.71828$ is Euler's number. If we multiply both sides of Euler's formula by the modulus, we get $|z|e^{i\theta} = |z|\cos\theta + |z|i\sin\theta$. The right-hand side of this equation is $x + iy$ (Chapter 7), such that $|z|e^{i\theta} = x + iy$. Any complex number expressed in the Cartesian form $z = x + iy$ can alternatively be expressed in the **exponential form** $z = |z|e^{i\theta}$. The exponential form combines the two components of polar form (the modulus $|z|$ and the argument θ) using Euler's number.

Note: When using exponential form or when using Euler's formula, the argument θ must be expressed in **radians** (**not** degrees). To convert an angle from degrees to radians, multiply by $\frac{\pi}{180}$. For example, $60°$ equates to $60\left(\frac{\pi}{180}\right) = \frac{\pi}{3}$ rad.

Why is Euler's formula true? In this chapter, we will demonstrate this by showing that $\cos\theta + i\sin\theta$ satisfies the properties of an exponential function $e^{i\theta}$. (A more direct way to prove Euler's formula will be shown in Chapter 12 using the Maclaurin series.) The **exponential function** $e^{i\theta}$ has the following properties:

- $e^0 = 1$. If you set $\theta = 0$ in $\cos\theta + i\sin\theta$, you get $\cos 0 + i\sin 0 = 1 + 0i = 1$.
- $e^{i\theta}e^{i\varphi} = e^{i(\theta+\varphi)}$. In terms of $\cos\theta + i\sin\theta$, this becomes
$$e^{i\theta}e^{i\varphi} = (\cos\theta + i\sin\theta)(\cos\varphi + i\sin\varphi)$$
$$= \cos\theta\cos\varphi + i\cos\theta\sin\varphi + i\sin\theta\cos\varphi + i^2\sin\theta\sin\varphi$$
$$= \cos\theta\cos\varphi + i^2\sin\theta\sin\varphi + i\cos\theta\sin\varphi + i\sin\theta\cos\varphi$$
$$= \cos\theta\cos\varphi - \sin\theta\sin\varphi + i(\cos\theta\sin\varphi + \sin\theta\cos\varphi)$$
$$= \cos(\theta+\varphi) + i\sin(\theta+\varphi) = e^{i(\theta+\varphi)}$$
The last line uses trig identities like $\cos(\theta+\varphi) = \cos\theta\cos\varphi - \sin\theta\sin\varphi$.
- $e^{-i\theta} = \frac{1}{e^{i\theta}}$. Cosine is an even function: $\cos(-\theta) = \cos\theta$. Sine is an odd function: $\sin(-\theta) = -\sin\theta$. Thus, $e^{-i\theta} = \cos(-\theta) + i\sin(-\theta) = \cos\theta - i\sin\theta$. Now we will show that this equals $\frac{1}{e^{i\theta}}$.
$$\frac{1}{e^{i\theta}} = \frac{1}{\cos\theta + i\sin\theta} = \frac{1}{\cos\theta + i\sin\theta}\left(\frac{\cos\theta - i\sin\theta}{\cos\theta - i\sin\theta}\right)$$

Chapter 11 – Euler's Formula

$$\frac{1}{e^{i\theta}} = \frac{\cos\theta - i\sin\theta}{\cos^2\theta + \sin^2\theta} = \frac{\cos\theta - i\sin\theta}{1} = \cos\theta - i\sin\theta = e^{-i\theta}$$

We used the trig identity $\cos^2\theta + \sin^2\theta = 1$.

- $e^{i\theta}e^{-i\varphi} = \frac{e^{i\theta}}{e^{i\varphi}} = e^{i\theta - i\varphi}$ is simply a combination of the previous bullet points.
- $\left(e^{i\theta}\right)^n = e^{in\theta}$ is **de Moivre's theorem** (Chapter 10):
$$\left(e^{i\theta}\right)^n = (\cos\theta + i\sin\theta)^n = \cos n\theta + i\sin n\theta = e^{in\theta}$$

All these properties are characteristic of exponential functions. The ordinary real exponential function e^x has the properties that $e^0 = 1$, $e^x e^y = e^{x+y}$, $e^{-x} = \frac{1}{e^x}$, $e^x e^{-y} = \frac{e^x}{e^y} = e^{x-y}$, and $(e^x)^n = e^{nx}$. We see that $e^{i\theta} = \cos\theta + i\sin\theta$ has these very same properties (apart from a periodicity issue that we will explore in Chapter 16). In Chapter 12, we'll use the Maclaurin series to discover a more direct way to see that $e^{i\theta} = \cos\theta + i\sin\theta$ is true.

Note that the **real and imaginary parts** of $e^{i\theta}$ are $\text{Re}(e^{i\theta}) = \cos\theta$ and $\text{Im}(e^{i\theta}) = \sin\theta$. The **complex conjugate** of $e^{i\theta}$ is $\overline{e^{i\theta}} = e^{-i\theta} = \cos\theta - i\sin\theta$ (see the third bullet point above).

The famous equation $e^{i\pi} + 1 = 0$ (which involves arguably the five most important numbers: 0, 1, e, π, and i) follows from Euler's formula. Set θ equal to π in Euler's formula to get $e^{i\pi} = \cos\pi + i\sin\pi = -1 + i(0) = -1$. (Note that π radians equates to 180°.)

Example 1. Evaluate $e^{i\pi/6}$.

Solution: Use Euler's formula. Compare $e^{i\pi/6}$ to $e^{i\theta}$ to see that $\theta = \frac{\pi}{6}$ radians. Note that $\frac{\pi}{6}$ radians equates to $\frac{\pi}{6}\frac{180°}{\pi} = 30°$. (To convert from radians to degrees, multiply by $\frac{180}{\pi}$.)

$$e^{i\pi/6} = \cos\left(\frac{\pi}{6}\right) + i\sin\left(\frac{\pi}{6}\right) = \frac{\sqrt{3}}{2} + \frac{i}{2}$$

Example 2. Express $i - 1$ in exponential form.

Solution: Compare $i - 1 = -1 + i$ with $z = x + iy$ to see that $x = -1$ and $y = 1$. Find the modulus and argument like we did in Chapter 7.

$$|z| = \sqrt{x^2 + y^2} = \sqrt{(-1)^2 + 1^2} = \sqrt{1+1} = \sqrt{2}$$

$$\theta = \tan^{-1}\left(\frac{y}{x}\right) = \tan^{-1}\left(\frac{1}{-1}\right) = \tan^{-1}(-1) = -45° + 180° = 135° = 135\frac{\pi}{180} = \frac{3\pi}{4}$$

Since $x < 0$ and $y > 0$, the answer for θ lies in Quadrant II. As discussed in Chapter 7, when $x < 0$, we must add 180° to the calculator's answer in order to put the angle in the right quadrant. **Don't** use degrees in exponential form; use radians instead. (To convert from degrees to radians, multiply by $\frac{\pi}{180}$.) Note that 135° equates to $\frac{3\pi}{4}$ **radians** (since $\frac{135}{180} = \frac{135 \div 45}{180 \div 45} = \frac{3}{4}$.) The exponential form of $i - 1$ is $|z|e^{i\theta} = \sqrt{2}e^{3i\pi/4}$. Since the trig functions are periodic, we may add $2i\pi k$ to the exponent where k is an integer. The answer is thus more generally expressed as $|z|e^{i\theta} = \sqrt{2}e^{3i\pi/4 + 2i\pi k}$.

Example 3. Express $6e^{5i\pi/6}$ in Cartesian form.

Solution: Compare $6e^{5i\pi/6}$ with $|z|e^{i\theta}$ to see that $|z| = 6$ and $\theta = \frac{5\pi}{6}$. Now convert the polar form to Cartesian form like we did in Chapter 7. If your calculator is in degrees mode, first convert $\frac{5\pi}{6}$ to degrees: $\frac{5\pi}{6}\frac{180°}{\pi} = \frac{900°}{6} = 150°$.

$$x = |z|\cos\theta = 6\cos 150° = 6\left(-\frac{\sqrt{3}}{2}\right) = -3\sqrt{3}$$

$$y = |z|\sin\theta = 6\sin 150° = 6\left(\frac{1}{2}\right) = 3$$

The Cartesian form of the complex number is $z = x + iy = -3\sqrt{3} + 3i$.

Chapter 11 Problems

Directions: Evaluate each expression.

(1) $e^{i\pi/3}$

(2) $e^{i\pi/4}$

(3) $e^{3i\pi/2}$

(4) $e^{2i\pi/3}$

(5) $e^{7i\pi/6}$

(6) $e^{-i\pi/3}$

(7) $e^{i\pi/12} e^{i\pi/4}$

(8) $\dfrac{e^{13i\pi/30}}{e^{3i\pi/5}}$

(9) $\left(e^{i\pi/15}\right)^5$

(10) $e^{5i\pi/4}\sqrt{2} - ie^{i\pi/3}$

(11) $\dfrac{\sqrt{2}+e^{i\pi/4}}{\sqrt{2}-e^{i\pi/4}}$

(12) $\left(i\sqrt{3} + e^{i\pi/3}\right)^2$

Directions: Express each complex number in exponential form.

(13) $7 + 7i$

(14) $\sqrt{2} - i\sqrt{6}$

(15) $27i$

(16) $i - \sqrt{3}$

(17) $\dfrac{5}{i}$

(18) $\overline{2 + 2i\sqrt{3}}$

(19) $(2 + 3i)(2 - 3i)$

(20) $\dfrac{15i - 9}{4 - i}$

(21) $\sqrt{-9i}$

Chapter 11 – Euler's Formula

Directions: Express each complex number in Cartesian form.

(22) $8e^{i\pi/6}$

(23) $10e^{2i\pi/3}$

(24) $\sqrt{3}e^{11i\pi/6}$

(25) $12e^{5i\pi/4}$

(26) $9e^{i\pi/2}$

(27) $7e^{8i\pi}$

(28) $\sqrt{6}e^{5i\pi/3}$

(29) $3e^{3i\pi}$

(30) $i\sqrt{2}e^{-i\pi/4}$

12 Complex Trigonometry

Here is something to think about. Recall that in Chapter 1 we began by introducing the imaginary number i as a way to solve equations like $x^2 = -1$. Would you expect that merely introducing the imaginary number i in this way would allow you to take the cosine of some angle and get an answer bigger than 1? For example, would i allow you to solve equations like $\cos \theta = 2$? (In the world of real numbers, $-1 \leq \cos \theta \leq 1$.) Think about this for a moment.

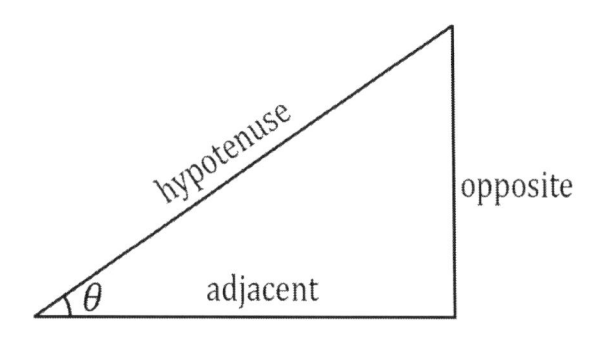

In trigonometry, the cosine function is defined in a right triangle relative to an angle as the side adjacent to the angle over the hypotenuse. The reason that the cosine function ordinarily can't exceed 1 is that in any real triangle, it isn't possible for either side to be larger than the hypotenuse.

The sine and cosine functions satisfy the identity $\sin^2 \theta + \cos^2 \theta = 1$, corresponding to the Pythagorean theorem. If we call the sides a, b, and c, where c is the hypotenuse, the Pythagorean theorem is $a^2 + b^2 = c^2$. In the world of real numbers, a^2 and b^2 are always nonnegative, but if we allow complex numbers, a^2 or b^2 may be negative. Now suppose that a and c are real, but that b is imaginary. In this case, $b^2 < 0$. When a negative b^2 is added to a^2, this makes c^2 smaller than a^2, meaning that the hypotenuse is smaller than one of the sides. If a is the side adjacent to angle θ, then $\frac{a}{c}$ would be larger than one, meaning that $\cos \theta > 1$. Thus, we see how simply introducing the idea that an imaginary number i solves the equation $x^2 = -1$ leads to the idea that the cosine of an angle could be greater than one.

Chapter 12 – Complex Trigonometry

It turns out that Euler's formula (Chapter 11) applies to **any** angle, even if the angle is **complex**. Below, we have written Euler's formula for θ and also for $-\theta$. The second formula is the complex conjugate of the first formula. Since cosine is an even function, $\cos(-\theta) = \cos\theta$, whereas since sine is an odd function, $\sin(-\theta) = -\sin\theta$. That's why the second equation below has positive cosine but negative sine.

$$e^{i\theta} = \cos\theta + i\sin\theta$$
$$e^{-i\theta} = \cos\theta - i\sin\theta$$

If we add these two equations together, we get the first equation below. If we subtract these two equations, we get the second equation below.

$$e^{i\theta} + e^{-i\theta} = 2\cos\theta$$
$$e^{i\theta} - e^{-i\theta} = 2i\sin\theta$$

If we divide by 2 in the first equation above, we get a formula for cosine. If we divide by $2i$ in the second equation above, we get a formula for sine.

$$\cos\theta = \frac{e^{i\theta} + e^{-i\theta}}{2}$$

$$\sin\theta = \frac{e^{i\theta} - e^{-i\theta}}{2i}$$

These formulas are valid for all values of θ, even imaginary ones (but you have to use radians; you can't use degrees). If the angle is purely imaginary, something fascinating happens. Suppose, for example, that $\theta = i\pi$. In this case, the formula for cosine gives:

$$\cos(i\pi) = \frac{e^{i(i\pi)} + e^{-i(i\pi)}}{2} = \frac{e^{i^2\pi} + e^{-i^2\pi}}{2} = \frac{e^{-\pi} + e^{-(-\pi)}}{2} = \frac{e^{-\pi} + e^{\pi}}{2} = \frac{e^{\pi} + e^{-\pi}}{2}$$

In the last step, we merely swapped the order of the addition (since $a + b = b + a$). If you know about the hyperbolic cosine and hyperbolic sine functions, you should recognize that this is the hyperbolic cosine of π. If you haven't heard of these functions before, don't worry. We'll introduce you to them now. The **hyperbolic** trig functions have an 'h' after them (like cosh θ instead of the usual cos θ) and are defined in terms of the exponential function as follows:

$$\cosh\theta = \frac{e^\theta + e^{-\theta}}{2}$$

$$\sinh\theta = \frac{e^\theta - e^{-\theta}}{2}$$

Analogous to the ordinary tangent function, the hyperbolic tangent equals hyperbolic sine divided by hyperbolic cosine: $\tanh\theta = \frac{\sinh\theta}{\cosh\theta}$.

Compare the hyperbolic sine and cosine functions with the formulas for sine and cosine that we obtained from Euler's formula. They are very similar. If you compare them closely, you should observe that:
$$\cos(i\theta) = \cosh\theta \quad , \quad \sin(i\theta) = i\sinh\theta$$
$$\cosh(i\theta) = \cos\theta \quad , \quad \sinh(i\theta) = i\sin\theta$$
You need to be careful when comparing the sine formulas since there is an i in the denominator of the formula for $\sin\theta$. Since $\frac{1}{i} = \frac{1}{i}\frac{i}{i} = \frac{i}{i^2} = \frac{i}{-1} = -i$, we may move the i from the denominator to the numerator if we also multiply by a negative sign. We will let you verify these formulas carefully in Example 7 and the exercises.

The above formulas are very handy. If you need to evaluate a trig function of a purely imaginary angle, these formulas allow you to instead evaluate a trig function of a purely real angle using the hyperbolic counterpart, as shown in the examples.

What if the angle is complex? For example, how would you find $\sin\left(\frac{\pi}{6} + \frac{i}{2}\right)$? For cases like this, use the **angle sum identities** below. See Example 4.
$$\cos(\theta + i\varphi) = \cos\theta\cosh\varphi - i\sin\theta\sinh\varphi$$
$$\sin(\theta + i\varphi) = \sin\theta\cosh\varphi + i\cos\theta\sinh\varphi$$
$$\cosh(\theta + i\varphi) = \cosh\theta\cos\varphi + i\sinh\theta\sin\varphi$$
$$\sinh(\theta + i\varphi) = \sinh\theta\cos\varphi + i\cosh\theta\sin\varphi$$
All cosines (ordinary or hyperbolic) are **even** functions, while all sine (ordinary or hyperbolic) are **odd** functions, as indicated below.
$$\cos(-\theta) = \cos\theta \quad , \quad \sin(-\theta) = -\sin\theta$$
$$\cosh(-\theta) = \cosh\theta \quad , \quad \sinh(-\theta) = -\sinh\theta$$

To really understand the relationship between sine, cosine, the exponential function, and hyperbolic trig functions, it helps to look at their Maclaurin series expansions. If you've taken a few semesters of calculus, you are probably familiar with these series. If not, don't worry. Although calculus is used to derive these series, we won't need to

Chapter 12 – Complex Trigonometry

use calculus to see how these different functions are related. The **Maclaurin[1] series** expansions for these functions are listed below. Note that θ must be expressed in **radians** (**not** degrees). The **factorial** notation means to multiply successively smaller integers until you reach one. For example, $4! = 4(3)(2)(1) = 24$.

$$\cos\theta = 1 - \frac{\theta^2}{2!} + \frac{\theta^4}{4!} - \frac{\theta^6}{6!} + \frac{\theta^8}{8!} - \frac{\theta^{10}}{10!} + \cdots$$

$$\sin\theta = \theta - \frac{\theta^3}{3!} + \frac{\theta^5}{5!} - \frac{\theta^7}{7!} + \frac{\theta^9}{9!} - \frac{\theta^{11}}{11!} + \cdots$$

$$e^\theta = 1 + \theta + \frac{\theta^2}{2!} + \frac{\theta^3}{3!} + \frac{\theta^4}{4!} + \frac{\theta^5}{5!} + \frac{\theta^6}{6!} + \frac{\theta^7}{7!} + \frac{\theta^8}{8!} + \frac{\theta^9}{9!} + \frac{\theta^{10}}{10!} + \frac{\theta^{11}}{11!} + \cdots$$

$$\cosh\theta = 1 + \frac{\theta^2}{2!} + \frac{\theta^4}{4!} + \frac{\theta^6}{6!} + \frac{\theta^8}{8!} + \frac{\theta^{10}}{10!} + \cdots$$

$$\sinh\theta = \theta + \frac{\theta^3}{3!} + \frac{\theta^5}{5!} + \frac{\theta^7}{7!} + \frac{\theta^9}{9!} + \frac{\theta^{11}}{11!} + \cdots$$

The Maclaurin series expansions for these functions show many close resemblances. They share similar powers and factorials. The main differences are whether they include odd terms, even terms, or both, and which terms are positive or negative. This aspect is very similar to the powers of i. Recall from Chapter 1 that $i^2 = -1$, $i^3 = -i$, $i^4 = 1$, $i^5 = i$, $i^6 = -1$, $i^7 = -i$, $i^8 = 1$, $i^9 = i$, $i^{10} = -1$, $i^{11} = -i$, etc. If we use these powers of i, we can see that Euler's formula, $e^{i\theta} = \cos\theta + i\sin\theta$, agrees with the first three series above.

$$e^{i\theta} = 1 + i\theta + \frac{(i\theta)^2}{2!} + \frac{(i\theta)^3}{3!} + \frac{(i\theta)^4}{4!} + \frac{(i\theta)^5}{5!} + \frac{(i\theta)^6}{6!} + \frac{(i\theta)^7}{7!} + \frac{(i\theta)^8}{8!} + \frac{(i\theta)^9}{9!} + \cdots$$

$$e^{i\theta} = 1 + i\theta + i^2\frac{\theta^2}{2!} + i^3\frac{\theta^3}{3!} + i^4\frac{\theta^4}{4!} + i^5\frac{\theta^5}{5!} + i^6\frac{\theta^6}{6!} + i^7\frac{\theta^7}{7!} + i^8\frac{\theta^8}{8!} + i^9\frac{\theta^9}{9!} + \cdots$$

$$e^{i\theta} = 1 + i\theta - \frac{\theta^2}{2!} - i\frac{\theta^3}{3!} + \frac{\theta^4}{4!} + i\frac{\theta^5}{5!} - \frac{\theta^6}{6!} - i\frac{\theta^7}{7!} + \frac{\theta^8}{8!} + i\frac{\theta^9}{9!} + \cdots$$

$$e^{i\theta} = \left(1 - \frac{\theta^2}{2!} + \frac{\theta^4}{4!} - \frac{\theta^6}{6!} + \frac{\theta^8}{8!} + \cdots\right) + i\left(\theta - \frac{\theta^3}{3!} + \frac{\theta^5}{5!} - \frac{\theta^7}{7!} + \frac{\theta^9}{9!} + \cdots\right)$$

$$e^{i\theta} = \cos\theta + i\sin\theta$$

---

[1] Maclaurin series are a special case of Taylor series, where the series is expanded about $x = 0$ instead of the more general $x = a$. That is, if you set $a = 0$ in a Taylor series expansion, you get the corresponding Maclaurin series expansion.

Complex Numbers Essentials Math Workbook with Answers

We can also see that $e^\theta = \cosh\theta + i\sinh\theta$ agrees with the last three series above. Similarly, we can see that $\cos(i\theta) = \cosh\theta$, $\sin(i\theta) = i\sinh\theta$, $\cosh(i\theta) = \cos\theta$, and $\sinh(i\theta) = i\sin\theta$.

To find an **inverse** function, note that:
$$\cos^{-1}(iz) = \pm i\cosh^{-1}z \quad , \quad \cosh^{-1}(iz) = \pm i\cos^{-1}z$$
$$\sin^{-1}(iz) = i\sinh^{-1}z \quad , \quad \sinh^{-1}(iz) = i\sin^{-1}z$$
$$\tan^{-1}(iz) = i\tanh^{-1}z \quad , \quad \tanh^{-1}(iz) = i\tan^{-1}z$$

For example, $\sinh^{-1}\left(\frac{\sqrt{3}}{2}i\right) = i\sin^{-1}\left(\frac{\sqrt{3}}{2}\right)$ has two principal answers: $\frac{\pi}{3}i$ or $\frac{2\pi}{3}i$ (since $\sin\frac{\pi}{3} = \sin\frac{2\pi}{3} = \frac{\sqrt{3}}{2}$). You can also add $2\pi ki$ (where k is an integer) to either answer.

The imaginary number i not only impacts trig functions, but also impacts **logarithms**. For example, consider $e^{i\pi} = -1$ (which we learned in Chapter 11). If we take the natural logarithm of both sides, we get $i\pi = \ln(-1)$. We used the property that $\ln e^u = u$. In the world of real numbers, you can't take the logarithm of a negative number. If you try finding $\ln(-1)$ on a calculator, you will get a domain error. But in the world of complex numbers, you can take the logarithm of a negative number; the answer is just imaginary. Since trig functions are periodic and $e^{i\pi} = \cos\pi + i\sin\pi$, we may add $2\pi ik$ to our answer, where k is an integer: $\ln(-1) = i\pi + 2\pi ik$. The following formula is very handy for finding natural logarithms in the context of complex numbers:
$$\ln z = \ln(|z|e^{i\theta}) = i\theta + 2\pi ik + \ln|z|$$
Here, k is an integer. In the case of $\ln(-1)$, $z = -1$, $|z| = 1$, and $\theta = \pi$. In this example, the formula gives $\ln(-1) = i\pi + 2\pi ik + \ln 1 = i\pi + 2\pi ik + 0 = i\pi + 2\pi ik$.

Example 1. Evaluate $\cos\left(\frac{\pi}{2}i\right)$ and $\sin\left(\frac{\pi}{2}i\right)$.

Solution: Use the formulas $\cos(i\theta) = \cosh\theta$ and $\sin(i\theta) = i\sinh\theta$ along with the definitions of $\sinh\theta$ and $\cosh\theta$.
$$\cos\left(\frac{\pi}{2}i\right) = \cosh\left(\frac{\pi}{2}\right) = \frac{e^{\pi/2} + e^{-\pi/2}}{2} \approx \frac{4.8105 + 0.2079}{2} \approx 2.51$$
$$\sin\left(\frac{\pi}{2}i\right) = i\sinh\left(\frac{\pi}{2}\right) = i\frac{e^{\pi/2} - e^{-\pi/2}}{2} \approx i\frac{4.8105 - 0.2079}{2} \approx 2.30i$$

Chapter 12 – Complex Trigonometry

It is interesting that $\cos\left(\frac{\pi}{2}i\right)$ is a real number, whereas $\sin\left(\frac{\pi}{2}i\right)$ is purely imaginary. Although $\cos\left(\frac{\pi}{2}i\right)$ is real, the answer (2.5) is outside the usual range of the cosine function. (As discussed at the beginning of the chapter, cosine is ordinarily less than 1.) In this case, cosine squared plus sine squared is $2.51^2 + (2.30i)^2 = 6.30 - 5.29 \approx 1$.

Example 2. Evaluate $\cosh\left(\frac{\pi}{6}i\right)$.

Solution: Use the formula $\cosh(i\theta) = \cos\theta$. Note that $\frac{\pi}{6}$ radians $= \frac{\pi}{6}\frac{180°}{\pi} = 30°$.

$$\cosh\left(\frac{\pi}{6}i\right) = \cos\left(\frac{\pi}{6}\right) = \frac{\sqrt{3}}{2}$$

The hyperbolic cosine function is ordinarily greater than one (when the argument is a real number). You can see this by looking at the definition $\cosh\theta = \frac{e^\theta + e^{-\theta}}{2}$. When θ is purely imaginary, we get a value for hyperbolic cosine that is less than one. (Note that $\frac{\sqrt{3}}{2} \approx 0.866$.)

Example 3. Evaluate $\tan(i\ln 3)$.

Solution: Find $\sin(i\ln 3)$ and $\cos(i\ln 3)$. Then divide to make tangent. Use the formulas $\sin(i\theta) = i\sinh\theta$ and $\cos(i\theta) = \cosh\theta$ and the definitions of $\sinh\theta$ and $\cosh\theta$. Note that $e^{\ln t} = t$ such that $e^{\ln 3} = 3$. Also, $-\ln u = \ln\frac{1}{u}$ such that $e^{-\ln 3} = e^{\ln 1/3} = \frac{1}{3}$.

$$\sin(i\ln 3) = i\sinh(\ln 3) = i\frac{e^{\ln 3} - e^{-\ln 3}}{2} = \frac{i}{2}\left(3 - \frac{1}{3}\right) = \frac{i}{2}\left(\frac{9}{3} - \frac{1}{3}\right) = \frac{i}{2}\frac{8}{3} = \frac{4i}{3}$$

$$\cos(i\ln 3) = \cosh(\ln 3) = \frac{e^{\ln 3} + e^{-\ln 3}}{2} = \frac{1}{2}\left(3 + \frac{1}{3}\right) = \frac{1}{2}\left(\frac{9}{3} + \frac{1}{3}\right) = \frac{1}{2}\frac{10}{3} = \frac{5}{3}$$

$$\tan(i\ln 3) = \frac{\sin(i\ln 3)}{\cos(i\ln 3)} = \frac{4i/3}{5/3} = \frac{4i}{3} \div \frac{5}{3} = \frac{4i}{3} \times \frac{3}{5} = \frac{4i}{5}$$

(Recall that the way to divide by a fraction is to multiply by its reciprocal.)

Example 4. Evaluate $\cos\left(\frac{\pi}{4} + i\ln 5\right)$.

Solution: Use the angle sum formula for cosine with a complex argument. Compare $\theta + i\varphi$ with $\frac{\pi}{4} + i\ln 5$ to identify $\theta = \frac{\pi}{4}$ and $\varphi = \ln 5$. Note that $\frac{\pi}{4}$ radians $= \frac{\pi}{4}\frac{180°}{\pi} = 45°$.

We will use the identity $e^{\ln t} = t$.

$$\cos(\theta + i\varphi) = \cos\theta \cosh\varphi - i \sin\theta \sinh\varphi$$

$$\cos\left(\frac{\pi}{4} + i\ln 5\right) = \cos\left(\frac{\pi}{4}\right)\cosh(\ln 5) - i\sin\left(\frac{\pi}{4}\right)\sinh(\ln 5)$$

$$\cos\left(\frac{\pi}{4} + i\ln 5\right) = \frac{\sqrt{2}}{2}\left(\frac{e^{\ln 5} + e^{-\ln 5}}{2}\right) - i\frac{\sqrt{2}}{2}\left(\frac{e^{\ln 5} - e^{-\ln 5}}{2}\right)$$

$$\cos\left(\frac{\pi}{4} + i\ln 5\right) = \frac{\sqrt{2}}{4}\left(e^{\ln 5} + e^{\ln 1/5}\right) - i\frac{\sqrt{2}}{4}\left(e^{\ln 5} - e^{\ln 1/5}\right)$$

$$\cos\left(\frac{\pi}{4} + i\ln 5\right) = \frac{\sqrt{2}}{4}\left(5 + \frac{1}{5}\right) - i\frac{\sqrt{2}}{4}\left(5 - \frac{1}{5}\right) = \frac{\sqrt{2}}{4}\left(\frac{26}{5}\right) - i\frac{\sqrt{2}}{4}\left(\frac{24}{5}\right)$$

$$\cos\left(\frac{\pi}{4} + i\ln 5\right) = \frac{\sqrt{2}}{2}\left(\frac{13}{5}\right) - i\frac{\sqrt{2}}{1}\left(\frac{12}{5}\right) = \frac{\sqrt{2}}{2}\left(\frac{13 - 12i}{5}\right) = \frac{13\sqrt{2}}{10} - \frac{6\sqrt{2}}{5}i$$

Example 5. Evaluate $\sinh^{-1}\left(\frac{i}{2}\right)$.

Solution: This is the **inverse** hyperbolic sine function. As usual in trig, the $^{-1}$ means inverse (it's **not** a reciprocal). Use the formula $\sinh^{-1}(iz) = i\sin^{-1}z$ to see that $\sinh^{-1}\left(\frac{i}{2}\right) = i\sin^{-1}\left(\frac{1}{2}\right)$. Which angle can you take the sine of and get $\frac{1}{2}$ as the answer? In Quadrants I or II (where sine is positive), the angles are $\frac{\pi}{6}$ or $\frac{5\pi}{6}$ (corresponding to 30° and 150°). We need to multiply by i according to $\sinh^{-1}\left(\frac{i}{2}\right) = i\sin^{-1}\left(\frac{1}{2}\right)$, which makes our final answers $\frac{\pi}{6}i + 2\pi ik$ or $\frac{5\pi}{6}i + 2\pi ik$, where k is an integer (adding any number of $2\pi i$'s gives an equivalent answer because the sine function is periodic). Check the answer: It's easy to make a mistake with an inverse function, but pretty straightforward to check the answer. We'll use the formula $\sinh(i\theta) = i\sin\theta$.

$$\sinh\left(\frac{\pi}{6}i\right) = i\sin\frac{\pi}{6} = i\left(\frac{1}{2}\right) = \frac{i}{2}$$

$$\sinh\left(\frac{5\pi}{6}i\right) = i\sin\frac{5\pi}{6} = i\left(\frac{1}{2}\right) = \frac{i}{2}$$

Example 6. Evaluate $\ln i$.

Solution: Use the formula $\ln z = \ln(|z|e^{i\theta}) = i\theta + 2\pi i k + \ln|z|$. Here, $z = i$, such that $|z| = 1$ and $\theta = \frac{\pi}{2}$ (recall Chapter 7). This gives $\ln i = \ln(1e^{i\pi/2}) = \frac{i\pi}{2} + 2\pi i k + \ln 1 = \frac{i\pi}{2} + 2\pi i k + 0 = \frac{i\pi}{2} + 2\pi i k$.

Check the answer: Exponentiate the answer to get $e^{i\pi/2 + 2\pi i k}$. Use the rule $e^{z+w} = e^z e^w$ to get $e^{i\pi/2 + 2\pi i k} = e^{i\pi/2} e^{2\pi i k}$. Use Euler's formula (Chapter 11) for each: $\left(\cos\frac{\pi}{2} + i\sin\frac{\pi}{2}\right)(\cos 2\pi k + i\sin 2\pi k) = (0 + 1i)(1 + 0i) = (1i)(1) = i$. Finally, take the natural logarithm of both sides of $e^{i\pi/2 + 2\pi i k} = i$ to get $\frac{i\pi}{2} + 2\pi i k = \ln(i)$.

Example 7. Derive $\cos(i\theta) = \cosh\theta$.

Solution: Replace θ with $i\theta$ in the formula for $\cos\theta$ in terms of exponentials.

$$\cos\theta = \frac{e^{i\theta} + e^{-i\theta}}{2} \rightarrow \cos(i\theta) = \frac{e^{i(i\theta)} + e^{-i(i\theta)}}{2} = \frac{e^{i^2\theta} + e^{-i^2\theta}}{2} = \frac{e^{-\theta} + e^{-(-1)\theta}}{2}$$

$$\cos(i\theta) = \frac{e^{-\theta} + e^{\theta}}{2} = \frac{e^{\theta} + e^{-\theta}}{2} = \cosh\theta$$

In the last step, we used the fact that addition is commutative: $a + b = b + a$.

Example 8. Derive $\cos(\theta + i\varphi) = \cos\theta\cosh\varphi - i\sin\theta\sinh\varphi$.

Solution: Begin with the usual angle sum identity for cosine from trigonometry, which is $\cos(\theta + \beta) = \cos\theta\cos\beta - \sin\theta\sin\beta$. Make the substitution $\beta = i\varphi$. Use the identities $\cos(i\varphi) = \cosh\varphi$ and $\sin(i\varphi) = i\sinh\varphi$.

$$\cos(\theta + \beta) = \cos\theta\cos\beta - \sin\theta\sin\beta = \cos\theta\cos(i\varphi) - \sin\theta\sin(i\varphi)$$
$$\cos(\theta + \beta) = \cos\theta\cosh\varphi - i\sin\theta\sinh\varphi$$

Chapter 12 Problems

Directions: Evaluate each expression.

(1) $\cos(i\pi)$

(2) $\sin(i)$

(3) $\cos\left(-\frac{\pi}{6}i\right)$

(4) $\sin(i \ln 2)$

(5) $\cosh 0$

(6) $\sinh 0$

(7) $\cosh 1$

(8) $\sinh 1$

(9) $\cosh\left(\frac{\pi}{3}i\right)$

(10) $\sinh\left(\frac{\pi}{2}i\right)$

(11) $\cosh(i\pi)$

(12) $\sinh\left(\frac{5\pi}{6}i\right)$

(13) $\cosh\left(\frac{7\pi}{6}i\right)$

(14) $\sinh\left(-\frac{\pi}{4}i\right)$

(15) $\tan(i)$

(16) $\tan(-i\ln 2)$

(17) $\tanh\left(\frac{\pi}{6}i\right)$

(18) $\tanh\left(-\frac{\pi}{4}i\right)$

(19) $\sin\left(\frac{\pi}{2}+\frac{i}{2}\right)$

(20) $\cos\left(\frac{\pi}{3}-i\ln 3\right)$

(21) $\sin\left(\frac{i}{3}-\frac{\pi}{6}\right)$

(22) $\cosh\left(\frac{1}{2} + i\pi\right)$

(23) $\sinh\left(\ln 4 + \frac{2\pi}{3}i\right)$

(24) $\cosh\left(\frac{\pi}{2}i - e\right)$

(25) $\cosh^{-1}\left(\frac{i}{2}\right)$

(26) $\sinh^{-1}\left(\frac{\sqrt{2}}{2}i\right)$

(27) $\ln\left(\frac{\sqrt{2}}{2} + \frac{\sqrt{2}}{2}i\right)$

Directions: Derive each expression.

(28) $\sin(i\theta) = i\sinh\theta$ (29) $\cosh(i\theta) = \cos\theta$ (30) $\sinh(i\theta) = i\sin\theta$

(31) $\sin(\theta + i\varphi) = \sin\theta\cosh\varphi + i\cos\theta\sinh\varphi$

(32) $\cosh(\theta + i\varphi) = \cosh\theta\cos\varphi + i\sinh\theta\sin\varphi$

(33) $\sinh(\theta + i\varphi) = \sinh\theta\cos\varphi + i\cosh\theta\sin\varphi$

13 Roots of Complex Numbers

The idea of complex numbers began by introducing the imaginary number i as a way to solve $x^2 = -1$, such that the square root of negative one, $\sqrt{-1}$, is $\pm i$.[1] Consider this question: What happens if you take the square root of the imaginary number, that is $\sqrt{i}$? If you've been paying attention, you'll remember that we found a technique for finding the square roots of complex numbers in Chapter 10 when we introduced de Moivre's theorem. In fact, $\sqrt{i}$ was an exercise in Chapter 10 (see Exercise 8). We will solve this problem here using a different approach.

Let's write i using exponential form (Chapter 11). Compare $x + iy$ with i to see that $x = 0$ and $y = 1$, such that $|z| = \sqrt{x^2 + y^2} = \sqrt{0^2 + 1^2} = 1$ and $\theta = \frac{\pi}{2}$ radians[2] (which corresponds to 90°). Thus, we may write i in the exponential form $|z|e^{i\theta} = 1e^{i\pi/2} = e^{i\pi/2}$. This is easy to check using Euler's formula: $e^{i\pi/2} = \cos\frac{\pi}{2} + i\sin\frac{\pi}{2} = 0 + i(1) = i$.

This means that $\sqrt{i} = \sqrt{e^{i\pi/2}}$. Recall from algebra that a square root equates to an exponent of one-half: $u^{1/2} = \sqrt{u}$. This gives $\sqrt{i} = \left(e^{i\pi/2}\right)^{1/2}$. Now recall from algebra the rule $(u^a)^b = u^{ab}$. This means that we should multiply the exponents $i\pi/2$ and $1/2$ together to make the exponent $\left(\frac{i\pi}{2}\right)\left(\frac{1}{2}\right) = \frac{i\pi}{4}$. Also, remember that when we take a square root we need to include a $\pm$ sign because there are two possible answers. For example, $(-3)^2 = 9$ and $3^2 = 9$ show that ± 3 are both square roots of 9. Putting this together, we have $\sqrt{i} = \pm e^{i\pi/4}$. Now we will apply Euler's formula to put the answer in Cartesian form: $\sqrt{i} = \pm\left(\cos\frac{\pi}{4} + i\sin\frac{\pi}{4}\right) = \pm\left(\frac{\sqrt{2}}{2} + \frac{\sqrt{2}}{2}i\right)$. The two solutions for $\sqrt{i}$ include $\frac{\sqrt{2}}{2} + \frac{\sqrt{2}}{2}i$ and $-\frac{\sqrt{2}}{2} - \frac{\sqrt{2}}{2}i$.

---

[1] Note that $(-i)^2 = (-1)^2 i^2 = (1)(-1) = -1$ as well as $i^2 = -1$, such that $x = -i$ and $x = i$ both solve $x^2 = -1$. As we discussed in a note at the end of Chapter 1, if you leave the $\pm$ out with square roots of negative numbers, it can cause some unnecessary headaches.

[2] Although $\frac{y}{x} = \frac{1}{0}$ would be undefined, as discussed in the solution to Exercise 4 of Chapter 7, the inverse tangent is $\frac{\pi}{2}$ (equivalent to 90°) because the point $(0, i)$ lies on the $+y$-axis in the complex plane.

Chapter 13 – Roots of Complex Numbers

It's easy to check the answer. Just multiply the answer by itself. We'll check this for the solution $\frac{\sqrt{2}}{2} + \frac{\sqrt{2}}{2}i$.

$$\left(\frac{\sqrt{2}}{2} + \frac{\sqrt{2}}{2}i\right)\left(\frac{\sqrt{2}}{2} + \frac{\sqrt{2}}{2}i\right) = \frac{\sqrt{2}}{2}\frac{\sqrt{2}}{2} + \frac{\sqrt{2}}{2}\frac{\sqrt{2}}{2}i + \frac{\sqrt{2}}{2}i\frac{\sqrt{2}}{2} + \frac{\sqrt{2}}{2}i\frac{\sqrt{2}}{2}i$$

$$= \frac{2}{4} + \frac{2}{4}i + \frac{2}{4}i + \frac{2}{4}i^2 = \frac{1}{2} + \frac{i}{2} + \frac{i}{2} + \frac{1}{2}(-1) = 0 + i = i \checkmark$$

We'll use this method (which is more intuitive than the method from Chapter 10) to find the roots of complex numbers, including cube roots, fourth roots, and other roots. Following is a description of the strategy, which is illustrated in the examples:

- The problem $\sqrt[n]{x + iy}$ is equivalent to $(x + iy)^{1/n}$, where $\sqrt[n]{}$ means the nth root (for example, $\sqrt[3]{}$ is a cube root). Identify x and y.
- Find $|z| = \sqrt{x^2 + y^2}$ and $\theta = \tan^{-1}\left(\frac{y}{x}\right)$ using the method from Chapters 7 and 11. This puts the problem in the exponential form $|z|^{1/n}e^{i\theta/n}$. Note that $|z|^{1/n}$ is equivalent to $\sqrt[n]{|z|}$. Recall from Chapter 7 that if $x < 0$, you need to add π radians (equivalent to 180°) to the calculator's answer for the inverse tangent. The angle must be in **radians** (**not** degrees).
- Use Euler's formula to write $|z|^{1/n}e^{i\theta/n}$ in Cartesian form. Call this $x_1 + iy_1$ (where the subscripts indicate that this corresponds to the first solution).
- There will be n solutions. One way to obtain the other solutions is to add $\frac{2\pi}{n}, \frac{4\pi}{n}, \frac{6\pi}{n}$, etc. to the argument and use Euler's formula to put each solution in Cartesian form. Another method is to use the formulas below:[3]

$$x_2 = x_1 \cos\left(\frac{2\pi}{n}\right) - y_1 \sin\left(\frac{2\pi}{n}\right) \quad , \quad x_3 = x_1 \cos\left(\frac{4\pi}{n}\right) - y_1 \sin\left(\frac{4\pi}{n}\right) \quad , \quad \text{etc.}$$

$$y_2 = x_1 \sin\left(\frac{2\pi}{n}\right) + y_1 \cos\left(\frac{2\pi}{n}\right) \quad , \quad y_3 = x_1 \sin\left(\frac{4\pi}{n}\right) + y_1 \cos\left(\frac{4\pi}{n}\right) \quad , \quad \text{etc.}$$

The second method can **sometimes** give exact answers in terms of square roots when the first method gives decimal approximations, as shown in Example 2.

---

[3] This method uses trigonometry to rotate the first solution $2\pi/n$ radians in the complex plane. This works because the n solutions form a regular n-sided polygon. See Examples 2-3.

Complex Numbers Essentials Math Workbook with Answers

- You can check each answer by multiplying it by itself n times. If the product equals $x + iy$, the answer checks out.

Example 1. Find all of the solutions to $\sqrt{3 + 4i}$.

Solution: Compare $3 + 4i$ with $z = x + iy$ to see that $x = 3$ and $y = 4$. Find the modulus and argument using the formulas from Chapter 7.

$$|z| = \sqrt{x^2 + y^2} = \sqrt{3^2 + 4^2} = \sqrt{9 + 16} = \sqrt{25} = 5$$

$$\theta = \tan^{-1}\left(\frac{y}{x}\right) = \tan^{-1}\left(\frac{4}{3}\right) \approx 53.1301° \times \frac{\pi}{180°} \approx 0.9273 \text{ rad (Quad. I)}$$

The problem $\sqrt{3 + 4i}$ is equivalent to $\sqrt{5e^{0.9273i}}$. In this problem, $n = 2$, since a square root is equivalent to a power of $1/2$. Thus, $\sqrt{5e^{0.9273i}}$ is equivalent to $5^{1/2}e^{0.9273i/2} \approx \sqrt{5}e^{0.4637i}$. Use Euler's formula:

$$\sqrt{5}e^{0.4637i} = \sqrt{5}\cos 0.4637 + i\sqrt{5}\sin 0.4637$$

Make sure that your calculator is in **radians** mode before using your calculator.[4]

$$\sqrt{5}e^{0.4637i} \approx 2 + 1i = 2 + i$$

That's one solution. Since $n = 2$, there is a second solution. To find the second solution, add $\frac{2\pi}{n} = \frac{2\pi}{2} = \pi$ radians to the argument, which is 0.4637, to get a new argument of $0.4637 + \pi \approx 3.605$. The second solution is

$$\sqrt{5}e^{3.605i} = \sqrt{5}\cos 3.605 + i\sqrt{5}\sin 3.605 \approx -2 - 1i = -2 - i$$

Since $\cos(\theta + \pi) = -\cos\theta$ and $\sin(\theta + \pi) = -\sin\theta$, for a **square root** the second solution will always be the negative of the first solution. For cube and other roots, it won't be as simple as reversing the sign. (You should recognize that if you follow the method from Chapter 10, the solution would be essentially the same.)

Check the answer: $(2 + i)(2 + i) = 4 + 2i + 2i - i^2 = 4 + 4i - 1 = 3 + 4i$ ✓
$(-2 - i)(-2 - i) = 4 + 2i + 2i - 1 = 4 + 4i - 1 = 3 + 4i$ ✓

---

[4] To find the cosine and sine, you could multiply 0.4637 radians by $\frac{180°}{\pi}$ to convert it to 26.57° and then use the calculator in degrees mode, but since Euler's formula itself and exponential form require the argument to be in radians, it's a good idea to prefer to work exclusively in radians for these calculations.

Chapter 13 – Roots of Complex Numbers

Example 2. Find all of the solutions to $\sqrt[3]{46i - 9}$.

Solution: Note that $46i - 9 = -9 + 46i$. Compare $-9 + 46i$ with $z = x + iy$ to see that $x = -9$ and $y = 46$. Find the modulus and argument using the formulas from Chapter 7.

$$|z| = \sqrt{x^2 + y^2} = \sqrt{(-9)^2 + 46^2} = \sqrt{81 + 2116} = \sqrt{2197} \approx 46.8722$$

$$\theta = \tan^{-1}\left(\frac{y}{x}\right) = \tan^{-1}\left(\frac{46}{-9}\right) \approx -1.3776 + \pi \approx 1.764 \text{ rad (Quad. II)}$$

As discussed in Chapter 7, if $x < 0$, we add π radians (equivalent to 180°)[5] to the calculator's answer for inverse tangent to place the angle in the correct quadrant. The problem $\sqrt[3]{46i - 9}$ is equivalent to $\sqrt[3]{46.8722 e^{1.764i}}$. Since $n = 3$, $\sqrt[3]{46.8722 e^{1.764i}}$ is equivalent to $46.8722^{1/3} e^{1.764i/3} = 46.8722^{1/3} e^{0.588i}$. Use Euler's formula:

$$46.8722^{1/3} e^{0.588i} = 46.8722^{1/3} \cos 0.588 + i 46.8722^{1/3} \sin 0.588$$

Make sure that your calculator is in **radians** mode before using your calculator.

$$46.8722^{1/3} e^{0.588i} \approx 3 + 2i$$

That's one solution. Since $n = 3$, there are three solutions. We will use two different methods to find the other solutions and compare.

- Method 1: Add increments of $\frac{2\pi}{n} = \frac{2\pi}{3}$ radians to the argument, which is 0.588, to get new arguments of $0.588 + \frac{2\pi}{3} \approx 2.682$ and $0.588 + 2\left(\frac{2\pi}{3}\right) \approx 4.777$. The other solutions are:

$$46.8722^{1/3} e^{2.682i} = 46.8722^{1/3} \cos 2.682 + i 46.8722^{1/3} \sin 2.682 \approx -3.234 + 1.598i$$
$$46.8722^{1/3} e^{4.777i} = 46.8722^{1/3} \cos 4.777 + i 46.8722^{1/3} \sin 4.777 \approx 0.2328 - 3.598i$$

  Note that the first method gives decimal approximations.

- Method 2: Use the following formulas. Since the first solution is $3 + 2i$, in this example $x_1 = 3$ and $y_1 = 2$. Note that $\frac{2\pi}{n} = \frac{2\pi}{3} = \frac{2\pi}{3} \frac{180°}{\pi} = 120°$.

$$x_2 = x_1 \cos\left(\frac{2\pi}{3}\right) - y_1 \sin\left(\frac{2\pi}{3}\right) = 3\left(-\frac{1}{2}\right) - 2\left(\frac{\sqrt{3}}{2}\right) = -\frac{3}{2} - \sqrt{3}$$

$$y_2 = x_1 \sin\left(\frac{2\pi}{3}\right) + y_1 \cos\left(\frac{2\pi}{3}\right) = 3\left(\frac{\sqrt{3}}{2}\right) + 2\left(-\frac{1}{2}\right) = \frac{3\sqrt{3}}{2} - 1$$

---

[5] A calculator gives -1.3776 radians (or $-78.9298°$ if it is in degrees mode) for the inverse tangent. Add π radians to get 1.764.

$$x_3 = x_1 \cos\left(\frac{4\pi}{3}\right) - y_1 \sin\left(\frac{4\pi}{3}\right) = 3\left(-\frac{1}{2}\right) - 2\left(-\frac{\sqrt{3}}{2}\right) = -\frac{3}{2} + \sqrt{3}$$

$$y_3 = x_1 \sin\left(\frac{4\pi}{3}\right) + y_1 \cos\left(\frac{4\pi}{3}\right) = 3\left(-\frac{\sqrt{3}}{2}\right) + 2\left(-\frac{1}{2}\right) = -\frac{3\sqrt{3}}{2} - 1$$

The second solution is $x_2 + iy_2 = -\frac{3}{2} - \sqrt{3} + \frac{3\sqrt{3}}{2}i - i$ and the third solution is $x_3 + iy_3 = -\frac{3}{2} + \sqrt{3} - \frac{3\sqrt{3}}{2}i - i$. The second method gives exact answers in this case. If you use a calculator, you can check that these answers are equivalent to the first method. For example, $-\frac{3}{2} - \sqrt{3} \approx -3.232$ and $\frac{3\sqrt{3}}{2} - 1 \approx 1.598$.

The second method rotates the first solution by $\frac{2\pi}{3}$ radians (equivalent to 120°) in the complex plane. We plotted the three solutions in the complex plane below.

- The three solutions form a regular polygon with $n = 3$ sides. (For $n = 3$, it is an equilateral triangle.)
- Each solution has the same modulus ($|z_1| = |z_2| = |z_3| \approx 46.8722^{1/3} \approx 3.6$). The three solutions are therefore equidistant from the origin and lie on a circle with a radius of ≈ 3.6.
- The arguments of the solutions are separated by $\frac{2\pi}{n} = \frac{2\pi}{3}$ rad (or 120°).

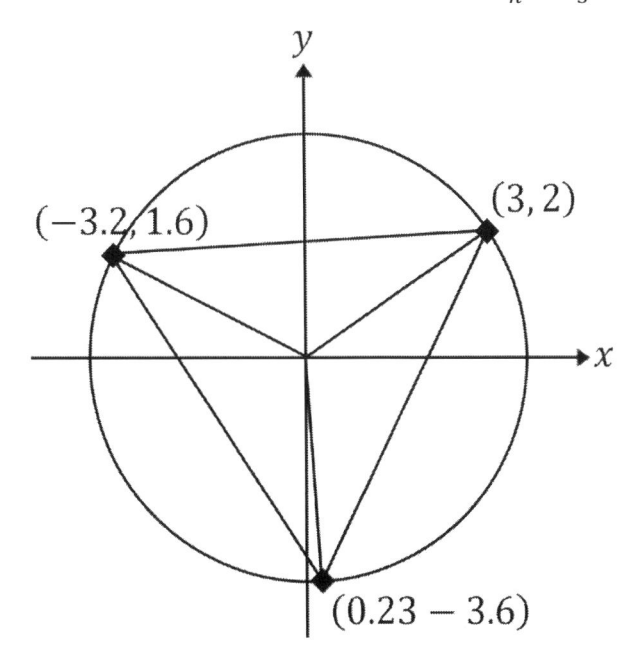

Check the answer: $(3 + 2i)(3 + 2i) = 3(3) + 3(2i) + 2i(3) + 2i(2i) = 9 + 6i + 6i - 4 = 5 + 12i$ such that $(3 + 2i)^3 = (3 + 2i)(3 + 2i)^2 = (3 + 2i)(5 + 12i) = 3(5) + 3(12i) + 2i(5) + 2i(12i) = 15 + 36i + 10i - 24 = -9 + 46i$ ✓
$(-3.234 + 1.598i)(-3.234 + 1.598i) \approx 10.459 - 5.168i - 5.168i - 2.554 = 7.905 - 10.336i$ and $(-3.234 + 1.598i)^3 \approx (7.905 - 10.336i)(-3.234 + 1.598i) \approx -25.565 + 12.632i + 33.427i - 16.517i^2 = -25.565 + 46.059i + 16.517 = -9.048 + 46.059i \approx -9 + 46i$ ✓
$(0.2328 - 3.598i)(0.2328 - 3.598i) \approx 0.054 - 0.838i - 0.838i - 12.946 = -12.892 - 1.676i$ and $(0.2328 - 3.598i)^3 \approx (-12.892 - 1.676i)(0.2328 - 3.598i) \approx -3.001 + 46.385i - 0.390i + 6.030i^2 = -3.001 + 45.995i - 6.030 = -9.031 + 45.995i \approx -9 + 46i$ ✓

Example 3. Find all of the solutions to $\sqrt[4]{28 - 96i}$.
Solution: Compare $28 - 96i$ with $z = x + iy$ to see that $x = 28$ and $y = -96$. Find the modulus and argument using the formulas from Chapter 7.
$$|z| = \sqrt{x^2 + y^2} = \sqrt{28^2 + (-96)^2} = \sqrt{784 + 9216} = \sqrt{10{,}000} = 100$$
$$\theta = \tan^{-1}\left(\frac{y}{x}\right) = \tan^{-1}\left(\frac{-96}{28}\right) \approx -73.7398 \times \frac{\pi}{180°} \approx -1.287 \text{ rad (Quad. IV)}$$
The problem $\sqrt[4]{28 - 96i}$ is equivalent to $\sqrt[4]{100e^{-1.287i}}$. Since $n = 4$, $\sqrt[4]{100e^{-1.287i}}$ is equivalent to $100^{1/4}e^{-1.287i/4} \approx 100^{1/4}e^{-0.3218i}$. Use Euler's formula:
$$100^{1/4}e^{-0.3218i} = 100^{1/4}\cos(-0.3218) + i100^{1/4}\sin(-0.3218)$$
Make sure that your calculator is in **radians** mode before using your calculator.
$$100^{1/4}e^{-0.3218i} \approx 3 - 1i = 3 - i$$
That's one solution. Since $n = 4$, there are four solutions.

- Method 1: Add increments of $\frac{2\pi}{n} = \frac{2\pi}{4} = \frac{\pi}{2}$ radians to the argument, which is -0.3218, to get new arguments of $-0.3218 + \frac{\pi}{2} \approx 1.249$, $-0.3218 + 2\left(\frac{\pi}{2}\right) \approx 2.820$, and $-0.3218 + 3\left(\frac{\pi}{2}\right) \approx 4.391$. The other solutions are:
$$100^{1/4}e^{-1.249i} = 100^{1/4}\cos 1.249 + i100^{1/4}\sin 1.249 \approx 1 + 3i$$
$$100^{1/4}e^{-2.820i} = 100^{1/4}\cos 2.820 + i100^{1/4}\sin 2.820 \approx -3 + i$$
$$100^{1/4}e^{-4.391i} = 100^{1/4}\cos 4.391 + i100^{1/4}\sin 4.391 \approx -1 - 3i$$

Complex Numbers Essentials Math Workbook with Answers

When all of the solutions happen to involve integers (not square roots), the first method is just as good as the second method, whereas Example 2 showed that if any solutions involve square roots, the second method can give the exact answers. The problem is that you often don't know before solving the problem whether the answers will involve square roots. This is why we prefer Method 2.

- Method 2: Use the following formulas. Since the first solution is $3 - i$, in this example $x_1 = 3$ and $y_1 = -1$. Note that $\frac{2\pi}{n} = \frac{2\pi}{4} = \frac{\pi}{2} = \frac{\pi}{2}\frac{180°}{\pi} = 90°$.

$$x_2 = x_1 \cos\left(\frac{2\pi}{4}\right) - y_1 \sin\left(\frac{2\pi}{4}\right) = 3(0) - (-1)(1) = 0 + 1 = 1$$

$$y_2 = x_1 \sin\left(\frac{2\pi}{4}\right) + y_1 \cos\left(\frac{2\pi}{4}\right) = 3(1) + (-1)(0) = 3 - 0 = 3$$

$$x_3 = x_1 \cos\left(\frac{4\pi}{4}\right) - y_1 \sin\left(\frac{4\pi}{4}\right) = 3(-1) - (-1)(0) = -3 + 0 = -3$$

$$y_3 = x_1 \sin\left(\frac{4\pi}{4}\right) + y_1 \cos\left(\frac{4\pi}{4}\right) = 3(0) + (-1)(-1) = 0 + 1 = 1$$

$$x_4 = x_1 \cos\left(\frac{6\pi}{4}\right) - y_1 \sin\left(\frac{6\pi}{4}\right) = 3(0) - (-1)(-1) = 0 - 1 = -1$$

$$y_4 = x_1 \sin\left(\frac{6\pi}{4}\right) + y_1 \cos\left(\frac{6\pi}{4}\right) = 3(-1) + (-1)(0) = -3 - 0 = -3$$

The second solution is $x_2 + iy_2 = 1 + 3i$, the third solution is $x_3 + iy_3 = -3 + i$, and the fourth solution is $x_4 + iy_4 = -1 - 3i$. Observe that these are the same answers as we obtained with Method 1.

The second method rotates the first solution by $\frac{\pi}{2}$ radians (equivalent to 90°) in the complex plane. We plotted the four solutions in the complex plane below.

- The four solutions form a regular polygon with $n = 4$ sides. (For $n = 4$, it is a square.)
- Each solution has the same modulus ($|z_1| = |z_2| = |z_3| = |z_4| \approx 100^{1/4} \approx 3.2$). The four solutions are therefore equidistant from the origin and lie on a circle with a radius of ≈ 3.2.
- The arguments of the solutions are separated by $\frac{2\pi}{n} = \frac{2\pi}{4} = \frac{\pi}{2}$ rad (or 90°). That is, look at the four lines connecting each point to the origin; these lines are all 90° apart.

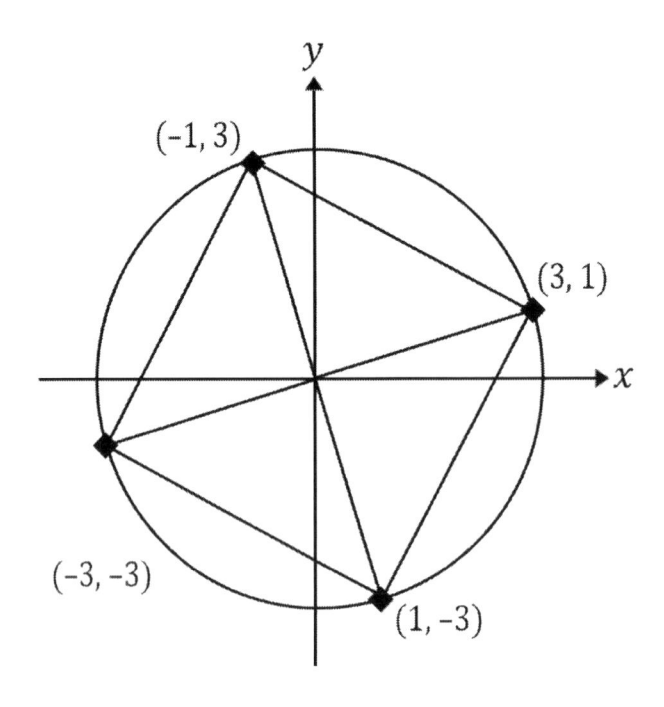

Check the answer: $(3-i)(3-i) = 9 - 3i - 3i + i^2 = 9 - 6i + (-1) = 8 - 6i$
$(3-i)^4 = (3-i)^2(3-i)^2 = (8-6i)(8-6i) = 8(8) + 8(-6i) - 6i(8) - 6i(-6i) = 64 - 48i - 48i + 36i^2 = 64 - 96i - 36 = 28 - 96i$ ✓
$(1+3i)(1+3i) = 1 + 3i + 3i + 9i^2 = 1 + 6i + (-9) = -8 + 6i$
$(1+3i)^4 = (1+3i)^2(1+3i)^2 = (-8+6i)(-8+6i) = -8(-8) - 8(6i) + 6i(-8) + 6i(6i) = 64 - 48i - 48i + 36i^2 = 64 - 96i - 36 = 28 - 96i$ ✓
$(-3+i)(-3+i) = 9 - 3i - 3i + i^2 = 9 - 6i + (-1) = 8 - 6i$
$(-3+i)^4 = (-3+i)^2(-3+i)^2 = (8-6i)(8-6i) = 8(8) + 8(-6i) - 6i(8) - 6i(-6i) = 64 - 48i - 48i + 36i^2 = 64 - 96i - 36 = 28 - 96i$ ✓
$(-1-3i)(-1-3i) = 1 + 3i + 3i + 9i^2 = 1 + 6i + (-9) = -8 + 6i$
$(-1-3i)^4 = (-1-3i)^2(-1-3i)^2 = (-8+6i)(-8+6i) = -8(-8) - 8(6i) + 6i(-8) + 6i(6i) = 64 - 48i - 48i + 36i^2 = 64 - 96i - 36 = 28 - 96i$ ✓

Chapter 13 Problems

Directions: Find all of the solutions (in Cartesian form) for each problem below.

(1) $\sqrt{65 - 72i}$

(2) $\sqrt{-3i}$

ns of Complex Numbers

(3) $\sqrt[3]{i-1}$

(4) $\sqrt[4]{-1+i\sqrt{3}}$

(5) $(65 - 142i)^{1/3}$

(6) $\sqrt[3]{i}$

Chapter 13 – Roots of Complex Numbers

(7) $\sqrt[4]{-\sqrt{2} - i\sqrt{6}}$

(8) $81^{1/4}$

(9) $\sqrt[3]{-64}$

(10) $(-208 - 144i)^{2/3}$

(11) $(-i)^{5/6}$

(12) $(8-6i)^{5/2}$

14 Roots of Unity

Consider the seemingly simple polynomial equation $z^n - 1 = 0$, where n is a positive integer. The solutions to this equation are referred to as the **roots of unity** because the solutions are the nth roots of one: $\sqrt[n]{1}$. This is a special case of the complex roots that we considered in Chapter 13.

It is easy to solve the above equation in exponential form. The first solution is 1 since $1^n - 1 = 0$. The remaining solutions are $e^{2i\pi/n}$, $e^{4i\pi/n}$, $e^{6i\pi/n}$, $e^{8i\pi/n}$, up to and including $e^{2(n-1)i\pi/n}$. To express the solutions in rectangular form, use Euler's formula (Chapter 11). For example, $e^{2i\pi/n} = \cos\frac{2\pi}{n} + i\sin\frac{2\pi}{n}$.

The following points apply to the roots of unity:
- There are n solutions to $z^n - 1 = 0$.
- Each solution has a modulus of one. The arguments are $0, \frac{2\pi}{n}, \frac{4\pi}{n}, \frac{6\pi}{n}$, etc. When plotted in the complex planes, the points are equally spaced around a unit circle (which has a radius of one), forming a **regular polygon**.
- If $n = 1, 2, 3, 4, 6,$ or 12, then $\frac{2\pi}{n} = 2\pi, \pi, \frac{2\pi}{3}, \frac{\pi}{2}, \frac{\pi}{3}$, or $\frac{\pi}{6}$ (corresponding to 360°, 180°, 120°, 90°, 60°, or 30°). For these cases, it is easy to use trigonometry to express the **exact solution** in rectangular form (since it is easy to find an exact value for the sine or cosine of these angles). For $n = 24$, for which $\frac{2\pi}{n} = \frac{\pi}{12}$ (or 15°), you can use the half-angle trig identities to find the exact solutions. (But if $n = 7$, for example, you would need the sine or the cosine of $\frac{360°}{7}$. For cases like this, in rectangular form you would just express the answer as a decimal.)
- If a solution to $z^n - 1 = 0$ is complex, its **complex conjugate** is also a solution. For example, if $\frac{1}{2} + \frac{\sqrt{3}}{2}i$ is a root, $\frac{1}{2} - \frac{\sqrt{3}}{2}i$ is also a root. Complex solutions to $z^n - 1 = 0$ come in **pairs**. Why? Put the polynomial equation in the form $z^n = 1$. The complex conjugate of both sides tells us that $\overline{z^n} = 1$, and the property $\overline{z^n} = \overline{z}^n$ (see Chapter 16) tells us that $\overline{z}^n = 1$ if $z^n = 1$. If z is a solution, so is $\overline{z}$.

Chapter 14 – Roots of Unity

- The polynomial equation $z^n - 1 = 0$ **factors** as

$$(z-1)\left(z - e^{\frac{2i\pi}{n}}\right)\left(z - e^{\frac{4i\pi}{n}}\right)\left(z - e^{\frac{6i\pi}{n}}\right) \cdots \left[z - e^{\frac{2(n-1)i\pi}{n}}\right] = 0$$

Note that $z = 1$, $z = e^{2i\pi/n}$, $z = e^{4i\pi/n}$ are each solutions to $z^n - 1 = 0$. Why? This multiplication makes a polynomial of degree n such that the polynomial equals zero for any value of z that is a solution to $z^n - 1 = 0$.

- The **sum** of all of the roots is zero if $n > 1$. That is,

$$1 + e^{\frac{2i\pi}{n}} + e^{\frac{4i\pi}{n}} + e^{\frac{6i\pi}{n}} + \cdots + e^{\frac{2(n-1)i\pi}{n}} = 0$$

We will rewrite this sum as follows:

$$1 + e^{1\left(\frac{2i\pi}{n}\right)} + e^{2\left(\frac{2i\pi}{n}\right)} + e^{3\left(\frac{2i\pi}{n}\right)} + \cdots + e^{(n-1)\left(\frac{2i\pi}{n}\right)} = 0$$

This is a **geometric series**. It is handy to know the geometric series formula,

$$1 + ar + ar^2 + ar^3 + \cdots + ar^{n-1} = \frac{a - ar^n}{1 - r}, \text{ which applies when } r \neq 1.$$ The sum of the roots is a geometric series with $a = 1$ and $r = e^{2i\pi/n}$, such that the sum is

$$\frac{a - ar^n}{1 - r} = \frac{1 - 1\left(e^{2i\pi/n}\right)^n}{1 - e^{2i\pi/n}} = \frac{1 - e^{2i\pi}}{1 - e^{2i\pi/n}} = \frac{1 - 1}{1 - e^{2i\pi/n}} = 0$$

We used Euler's formula for the last step: $e^{2i\pi} = \cos 2\pi + i \sin 2\pi = 1 + 0i = 1$. (Note that the sum **isn't** zero for the case $n = 1$. In this case, $r = e^{2i\pi/1} = 1$, and the geometric series formula doesn't apply when $r = 1$. The case $n = 1$ is trivial. The equation $z^1 - 1 = 0$ has one root: $z = 1$.)

The solutions to the equation $z^n - 1 = 0$, which are the **roots of unity**, are related to the solutions to the equation $z^n - u = 0$, which are the roots of the number u (since $u^{1/n} = \sqrt[n]{u}$ solves the equation $z^n - u = 0$). Note that u may be a complex number.

- The first solution is $\sqrt[n]{|u|}e^{i\theta/n}$, where $|u|$ and θ are the modulus and argument of the number u.
- A second solution is $\sqrt[n]{|u|}e^{i\theta/n}e^{2i\pi/n}$, where $e^{2i\pi/n}$ is a root of unity.
- A third solution is $\sqrt[n]{|u|}e^{i\theta/n}e^{4i\pi/n}$, where $e^{4i\pi/n}$ is another root of unity.
- The nth solution is $\sqrt[n]{|u|}e^{i\theta/n}e^{2(n-1)i\pi/n}$, where $e^{2(n-1)i\pi/n}$ is the last root of unity.

These solutions agree with the method that we used in Chapter 13 to find $\sqrt[n]{u}$.

Complex Numbers Essentials Math Workbook with Answers

Example 1. Express all of the solutions to $z^3 - 1 = 0$ in rectangular form.

Solution: Since z^3 has an exponent of 3, there are $n = 3$ solutions. The first solution is $z_1 = 1$ since $1^3 - 1 = 0$. The second solution is $z_2 = e^{2i\pi/3} = \cos\left(\frac{2\pi}{3}\right) + i\sin\left(\frac{2\pi}{3}\right) = -\frac{1}{2} + \frac{\sqrt{3}}{2}i$. The third solution is $z_3 = e^{4i\pi/3} = \cos\left(\frac{4\pi}{3}\right) + i\sin\left(\frac{4\pi}{3}\right) = -\frac{1}{2} - \frac{\sqrt{3}}{2}i$.

Notes: As expected, the complex solutions are a pair of complex conjugates. That is, $-\frac{1}{2} - \frac{\sqrt{3}}{2}i$ is the complex conjugate of $-\frac{1}{2} + \frac{\sqrt{3}}{2}i$, such that $z_3 = \overline{z_2}$. Also, the sum of the roots equals zero:

$$z_1 + z_2 + z_3 = 1 + \frac{1}{2} + \frac{\sqrt{3}}{2}i - \frac{1}{2} - \frac{\sqrt{3}}{2}i = 1 - \frac{1}{2} - \frac{1}{2} = 0$$

Check the answers: Verify that the cube of each solution equals one. The first solution is trivial: $1^3 = 1$. ✓

For the second solution, $\left(-\frac{1}{2} + \frac{\sqrt{3}}{2}i\right)^2 = \frac{1}{4} - \frac{\sqrt{3}}{4}i - \frac{\sqrt{3}}{4}i + \frac{3}{4}i^2 = \frac{1}{4} - \frac{\sqrt{3}}{2}i - \frac{3}{4} = -\frac{1}{2} - \frac{\sqrt{3}}{2}i$

$\left(-\frac{1}{2} + \frac{\sqrt{3}}{2}i\right)^3 = \left(-\frac{1}{2} + \frac{\sqrt{3}}{2}i\right)\left(-\frac{1}{2} + \frac{\sqrt{3}}{2}i\right)^2 = \left(-\frac{1}{2} + \frac{\sqrt{3}}{2}i\right)\left(-\frac{1}{2} - \frac{\sqrt{3}}{2}i\right)$

$= \frac{1}{4} + \frac{\sqrt{3}}{4}i - \frac{\sqrt{3}}{4}i - \frac{3}{4}i^2 = \frac{1}{4} - \frac{3}{4}(-1) = \frac{1}{4} + \frac{3}{4} = 1$ ✓

For the third solution, $\left(-\frac{1}{2} - \frac{\sqrt{3}}{2}i\right)^2 = \frac{1}{4} + \frac{\sqrt{3}}{4}i + \frac{\sqrt{3}}{4}i + \frac{3}{4}i^2 = \frac{1}{4} + \frac{\sqrt{3}}{2}i - \frac{3}{4} = -\frac{1}{2} + \frac{\sqrt{3}}{2}i$

$\left(-\frac{1}{2} - \frac{\sqrt{3}}{2}i\right)^3 = \left(-\frac{1}{2} - \frac{\sqrt{3}}{2}i\right)\left(-\frac{1}{2} - \frac{\sqrt{3}}{2}i\right)^2 = \left(-\frac{1}{2} - \frac{\sqrt{3}}{2}i\right)\left(-\frac{1}{2} + \frac{\sqrt{3}}{2}i\right)$

$= \frac{1}{4} - \frac{\sqrt{3}}{4}i + \frac{\sqrt{3}}{4}i - \frac{3}{4}i^2 = \frac{1}{4} - \frac{3}{4}(-1) = \frac{1}{4} + \frac{3}{4} = 1$ ✓

Chapter 14 Problems

Directions: Express all of the solutions to each equation in rectangular form.

(1) $z^2 - 1 = 0$

(2) $z^4 - 1 = 0$

(3) $z^6 - 1 = 0$

(4) $z^{12} - 1 = 0$

15 Roots of Polynomials

A **polynomial** consists of terms that are simple powers with coefficients (like $3x^5$ or $7x^2$) added (or subtracted) together. The general form of a polynomial is:
$$p(z) = a_n z^n + a_{n-1} z^{n-1} + a_{n-2} z^{n-2} + \cdots + a_3 x^3 + a_2 x^2 + a_1 x + a_0$$

- The **degree** of the polynomial refers to the highest power. In the equation above, the degree is n since the highest power appears in z^n. As another example, the polynomial $2x^4 - x^3 + 5x - 7$ has degree 4. Note that $n \geq 1$.
- The constants $a_n, a_{n-1}, a_{n-2}, \cdots, a_3, a_2, a_1$ are **coefficients**. (Technically, a_0 is the coefficient of x^0; recall from algebra that $x^0 = 1$, such that $a_0 x^0 = a_0$.)
- The term a_0 is a constant term. For example, in $3x^2 - 7x + 5$, the constant term is 5, whereas the polynomial $5x^4 - 7x$ doesn't have a constant term.
- The term $a_1 x$ is a linear term, the term $a_2 x^2$ is a quadratic term, the term $a_3 x^3$ is a cubic term, the term $a_4 x^4$ is a quartic term, etc.

If $p(c) = 0$, then that constant c is called a **root** (or a **zero**) of the polynomial $p(z)$; the value c is a **solution** to the equation $p(z) = 0$. A polynomial of degree n has at most n **distinct** roots; it will have fewer than n distinct roots if there is a **double** (or triple, quadruple, etc.) root. (Recall that we encountered double roots in the context of quadratic equations in Chapter 9.) When there are double roots, triple roots, etc., the **multiplicity** refers to the number of identical roots; for example, the multiplicity is 2 for a double root and 3 for a triple root.

Polynomials with real coefficients **don't** always have real solutions. As noted in Chapter 1, the seemingly simple polynomial $x^2 + 1$ only has **imaginary** roots (since the solutions to $x^2 + 1 = 0$ are $x = \pm i$). In general, some (or all) of the roots may be **complex**. In this chapter, we will discuss a few important theorems concerning polynomials and the nature of the roots.

An alternative way to express a general polynomial is to use the n roots as follows:
$$p(z) = a_n(z - c_1)(z - c_2)(z - c_3) \cdots (z - c_n)$$
where $c_1, c_2, c_3, \cdots, c_n$ are the n **roots** of the polynomial $p(z)$.

If you multiply the previous equation out for small values of n, you will see a pattern:
$$a_2(z-c_1)(z-c_2) = a_2z^2 - a_2(c_1+c_2)z + a_2c_1c_2$$
$$a_3(z-c_1)(z-c_2)(z-c_3)$$
$$= a_3z^3 - a_3(c_1+c_2+c_3)z^2 + a_3(c_1c_2+c_1c_3+c_2c_3)z - a_3c_1c_2c_3$$
$$a_4(z-c_1)(z-c_2)(z-c_3)(z-c_4)$$
$$= a_4z^4 - a_4(c_1+c_2+c_3+c_4)z^3$$
$$+ a_4(c_1c_2+c_1c_3+c_1c_4+c_2c_3+c_2c_4+c_3c_4)z^2$$
$$- (c_1c_2c_3+c_1c_2c_4+c_1c_3c_4+c_2c_3c_4)z + a_4c_1c_2c_3c_4$$

The equations above form the following pattern for the general polynomial:
- The leading term is $a_n z^n$.
- The signs of the terms alternate; the second term is negative, the third term is positive, the fourth term is negative, etc.
- The coefficient of z^{n-1} is $-a_n$ times the sum of the roots: $c_1 + c_2 + c_3 + \cdots + c_n$.
- The coefficient of z^{n-2} is a_n times the sum all possible pairs of the roots: $c_1c_2 + c_1c_3 + c_1c_4 + \cdots + c_{n-1}c_n$.
- The coefficient of z^{n-3} is $-a_n$ times the sum all possible triplets of the roots: $c_1c_2c_3 + c_1c_2c_4 + c_1c_2c_5 + \cdots + c_{n-2}c_{n-1}c_n$.
- The constant term is a_n (or $-a_n$) times the product of the roots: $c_1c_2c_3 \cdots c_n$.

As an example, consider the polynomial with roots $c_1 = 2$, $c_2 = 3$, and $c_3 = 5$, where $a_3 = 1$. This polynomial factors as $(1)(z-2)(z-3)(z-5) = (z-2)(z-3)(z-5)$. If we multiply this out, we get
$$(z-2)(z^2 - 5z - 3z + 15) = (z-2)(z^2 - 8z + 15)$$
$$= z^3 - 8z^2 + 15z - 2z^2 + 16z - 30 = z^3 - 10z^2 + 31z - 30$$
Note that the forms $z^3 - 10z^2 + 31z - 30$ and $(z-2)(z-3)(z-5)$ are equivalent. This is a cubic polynomial with coefficients $a_3 = 1$, $a_2 = -10$, $a_1 = 31$, and $a_0 = -30$. Observe that
$$-a_3(c_1+c_2+c_3) = -(1)(2+3+5) = -10$$
$$a_3(c_1c_2+c_1c_3+c_2c_3) = (1)[(2)(3)+(2)(5)+(3)(5)] = 6+10+15 = 31$$
$$-a_3c_1c_2c_3 = -(1)(2)(3)(5) = -30$$
This shows that the formula $a_3(z-c_1)(z-c_2)(z-c_3) = a_3z^3 - a_3(c_1+c_2+c_3)z^2 + a_3(c_1c_2+c_1c_3+c_2c_3)z - a_3c_1c_2c_3$, which was given above, works for this example.

Chapter 15 – Roots of Polynomials

If we compare the general form of a polynomial in terms of $a_n, a_{n-1}, a_{n-2}, \cdots, a_0$ to the factored form in terms of the roots $c_1, c_2, c_3, \cdots, c_n$, which we multiplied out for the cases $n = 2$ thru 4, we obtain **Vieta's formulas** below.

$$a_{n-1} = -a_n(c_1 + c_2 + c_3 + \cdots + c_n)$$
$$a_{n-2} = a_n(c_1c_2 + c_1c_3 + c_1c_4 + \cdots + c_{n-1}c_n)$$
$$a_{n-3} = -a_n(c_1c_2c_3 + c_1c_2c_4 + c_1c_2c_5 + \cdots + c_{n-2}c_{n-1}c_n)$$
$$\vdots$$
$$a_0 = (-1)^n a_n c_1 c_2 c_3 \cdots c_n$$

The sum for a_{n-2} includes all possible pairs of the roots, the sum for a_{n-3} includes all possible triplets of the roots, etc. Note that $(-1)^n$ determines the sign of the constant term; the constant term is positive if n is even and negative if n is odd.

Any polynomial with degree 4 or lower can always be solved exactly using algebra:
- If $n = 1$, we get **linear** polynomial: $a_1 z + a_0$. The only root is $c_1 = -\frac{a_0}{a_1}$ (which is found by setting $a_1 z + a_0$ equal to zero).
- If $n = 2$, we get a quadratic: $a_2 z^2 + a_1 z + a_0$. The roots can be found by factoring or using the **quadratic formula** (recall Chapter 9) with $a_2 z^2 + a_1 z + a_0 = 0$.
$$z = \frac{-a_1 \pm \sqrt{a_1^2 - 4a_2 a_0}}{2a_2}$$
- If $n = 3$, we get a cubic: $a_3 z^3 + a_2 z^2 + a_1 z + a_0$. Later in this chapter, we will outline a strategy that helps to **factor** a polynomial. If a cubic doesn't factor easily, you can find a **cubic formula** in the Appendix, but beware that the cubic and quartic formulas are much more involved than the quadratic formula.
- If $n = 4$, we get a quartic: $a_4 z^4 + a_3 z^3 + a_2 z^2 + a_1 z + a_0$. There is a **quartic formula** (see the Appendix), though if you can factor a quartic, that's simpler.

If $n \geq 5$, there **isn't** a magic formula that solves the most general case.[1] However, there are techniques that allow you to find roots for a variety of cases. We will outline a strategy for finding roots and factoring polynomials after we discuss a few important theorems regarding polynomials.

---
[1] There do exist numerical techniques for finding roots of polynomials to a good decimal approximation.

If we multiply the $(z - c_i)$'s together for each root c_i of the polynomial $p(z)$, this makes the polynomial:
$$p(z) = a_n(z - c_1)(z - c_2)(z - c_3) \cdots (z - c_n)$$
If we divide $p(z)$ by any of these $(z - c_i)$'s, it makes a polynomial with lesser degree. For example, the following division results in a polynomial with degree $n - 1$.
$$\frac{p(z)}{z - c_1} = a_n(z - c_2)(z - c_3) \cdots (z - c_n)$$
If we consider dividing the polynomial $p(z)$ by the quantity $z - k$, where the constant k isn't necessarily one of the roots, we can learn something about the nature of the remainder. Before we think about polynomial division, let's briefly recall the division of ordinary arithmetic. If we divide 23 by 5, we get 4 with a remainder of 3, which we can express as $\frac{23}{5} = 4 + \frac{3}{5}$. (You can verify with a calculator that both sides equal 4.6); 23 is the **dividend**, 5 is the **divisor**, 4 is the **quotient**, and 3 is the **remainder**. We can express the polynomial division $\frac{p(z)}{z-k}$ similarly as shown below; $p(z)$ is the **dividend**, $z - k$ is the **divisor**, $q(z)$ is the **quotient**, and $r(z)$ is the **remainder**.
$$\frac{p(z)}{z - k} = q(z) + \frac{r(z)}{z - k}$$
According to the **remainder theorem**, if the polynomial $p(z)$ is divided by the quantity $z - k$ (where k is a constant), the remainder is equal to $p(k)$. Note that $p(k)$ means to evaluate the polynomial at k.[2] That is, $r(z) = p(k)$
$$\frac{p(z)}{z - k} = q(z) + \frac{p(k)}{z - k}$$
Why is the remainder theorem true? That is, why does $r(z) = p(k)$. First, $r(z)$ must be a constant because the remainder of polynomial division always has a lower degree than the divisor. For example, if you divide $36z^5 + 81z^4 - 2z^3 + 36z^2$ by $9z^2 + 4$, the remainder must have a degree lower than 2 (so it would either be proportional to the first power of z or it would be a constant). In the remainder theorem, the divisor is $z - k$, so in order for the remainder to have a lower degree, it must be a constant. Let's call that constant h, meaning that $r(z) = h$. With this notation, the polynomial division

---

[2] This means to plug k into the polynomial in place of z and do arithmetic to get a numerical value. For example, if $p(z) = 3z^3 - 4z + 2$, then $p(2) = 3(2)^3 - 4(2) + 2 = 3(8) - 8 + 2 = 24 - 6 = 18$.

becomes $\frac{p(z)}{z-k} = q(z) + \frac{h}{z-k}$. Multiply by $z - k$ on both sides of this equation to get $p(z) = q(z)(z - k) + h$. If we evaluate $p(z)$ at the value $z = k$, this equation becomes $p(k) = q(k)(k - k) + h$, which simplifies to $p(k) = h$, which equals the remainder. This shows that $p(k), r(z), h$, and k are all the same; in particular, since $r(z) = p(k)$, the remainder theorem is true.

According to the **factor theorem**, if polynomial $p(z)$ is evenly divisible by $z - c$, then c is a **root** of $p(z)$, which means that $p(c) = 0$. The factor theorem follows from the remainder theorem. In the remainder theorem, for any constant k (whether or not it is a root), the division $\frac{p(z)}{z-k}$ has a remainder equal to $p(k)$. Now consider the special case $k = c$ (meaning that the constant is a root). In this case, $\frac{p(z)}{z-c}$ has a remainder of $p(c)$. If c is a root of $p(z)$, then $p(c) = 0$. (That was how we defined the root of a polynomial in the beginning of this chapter.)

According to the **fundamental theorem of algebra**, if polynomial $p(z)$ has degree n, then $p(z)$ has n roots; in the most general case, some of these roots will be **complex numbers**, and in some cases some of these roots may be identical (with **multiplicity** equal to 2 or more). If you think about the factored form of $p(z)$ that we used to obtain Vieta's formulas, this should make sense. You need to multiply n terms of the form $(z - c)$ together in order to make a term with the highest power z^c.

Even if all of the coefficients of $p(z)$ are real, some of the roots may be **complex**. For example, consider the polynomial $z^4 + 2z^2 + 1$, which has real coefficients $a_4 = 1$, $a_3 = 0, a_2 = 2, a_1 = 0$, and $a_0 = 1$. For any real value of z, $z^4 + 2z^2 + 1$ is greater than or equal to 1 because z^2 and z^4 are each positive if z is real. The only way $z^4 + 2z^2 + 1$ can be zero is if z is complex. In fact, since $z^4 + 2z^2 + 1$ factors as $(z^2 + 1)(z^2 + 1)$, the roots of $z^4 + 2z^2 + 1$ are $z = -i$ and $z = i$. This is easy to check: $i^4 + 2i^2 + 1 = 1 - 2 + 1 = 0$. In this example, $z = -i$ and $z = i$ each have a multiplicity of 2 (that is, both are double roots).

If all of the coefficients of $p(z)$ are real, if c is a root of $p(z)$, then its **complex conjugate**, $\overline{c}$, is also a root of $p(z)$. That is, if $p(c) = 0$, then $p(\overline{c}) = 0$. Put another way, when all

of the coefficients of a polynomial are real, any complex roots will come in **pairs** (where the two complex numbers are **complex conjugates** of one another). The reason for this is that any number times its complex conjugate, $c\bar{c}$, is always real (Chapter 3). Think about the factored form of the polynomial:
$$p(z) = a_n(z - c_1)(z - c_2)(z - c_3) \cdots (z - c_n)$$
Suppose that $c_1 = 4 + 5i$ is a complex root. How could $(z - 4 - 5i)$ multiply other factors and result in real coefficients in the polynomial $p(z)$? The answer is that one of the other factors must involve its complex conjugate: $c_2 = 4 - 5i$. When these two factors are multiplied together, they make real coefficients. Watch how $5iz$ and $20iz$ cancel out in the multiplication:
$$(z - 4 - 5i)(z - 4 + 5i) = z^2 - 4z + 5iz - 4z + 16 - 20i - 5iz + 20i - 25i^2$$
$$= z^2 - 8z + 16 - 25i^2 = z^2 - 8z + 16 + 25 = z^2 - 8z + 41$$

Similarly, if all of the coefficients of $p(z)$ are rational, if any of the roots of $p(z)$ include **square roots**, like $2 + \sqrt{3}$, then its conjugate is also a root. The conjugate of $2 + \sqrt{3}$ is $2 - \sqrt{3}$. If all of the coefficients of $p(z)$ are rational, any roots with **square roots** will come in **pairs** (where the two irrational numbers are conjugates of one another). When the conjugate pairs are multiplied, the coefficients are rational:
$$(z - 2 - \sqrt{3})(z - 2 + \sqrt{3}) = z^2 - 2z + z\sqrt{3} - 2z + 4 - 2\sqrt{3} - z\sqrt{3} + 2\sqrt{3} - 3$$
$$= z^2 - 4z + 1$$

If all of the coefficients of $p(z)$ are **rational**, we may multiply the equation $p(z) = 0$ by the least common denominator to make a new equation where every coefficient is an **integer**. For example, multiply by 6 on both sides of $\frac{z^2}{3} - \frac{3z}{2} + \frac{4}{3} = 0$ to get the equation $2z^2 - 9z + 8 = 0$. This offers a way to deal with purely integer coefficients when **fractions** are present.

If all of the coefficients of $p(z)$ are **integers**, if the root c is an integer, then the constant a_0 is evenly divisible by c. This fact helps to narrow the search for integer roots. For example, consider $z^3 - 3z^2 - 4z + 12$, where the roots include -2, 2, and 3. (You can check this by plugging each value into the polynomial.) Observe that the roots -2, 2, and 3 each evenly divide into 12.

Chapter 15 – Roots of Polynomials

If all of the coefficients of $p(z)$ are **integers**, if the root $c = \frac{N}{D}$ is a **reduced fraction**,[1] the leading coefficient a_n must be evenly divisible by D and the constant a_0 must be evenly divisible by N. For example, $4z^2 - 11z + 6$ has a leading coefficient of $a_2 = 4$ and a constant of $a_0 = 6$. This polynomial factors as $(4z - 3)(z - 2)$, so its roots are $z = \frac{3}{4}$ and $z = 2$. The root $z = \frac{3}{4}$ is a reduced fraction with $N = 3$ and $D = 4$. Observe that $a_2 = 4$ is evenly divisible by $D = 4$, while $a_0 = 6$ is evenly divisible by $N = 3$.

The following strategy is helpful for factoring and/or finding the roots of a large variety of polynomials, including the exercises at the end of this chapter. If you don't like this strategy or if you prefer something more direct, for cubics or quartics you may use the formula in the Appendix (but beware that the cubic and quartic formulas are not as simple as the quadratic formula; sometimes the solution can be quite tedious).

- If the degree is one, simply set the polynomial equal to zero and solve for the variable. For example, for $2z - 8$, we get $2z - 8 = 0$ for which $z = \frac{8}{2} = 4$.
- If the degree is two, it's a quadratic. See Chapter 9.
- If the degree is three or higher, **once you find one root, you can factor it out of the polynomial to make a simpler polynomial**. For example, if you determine that one of the roots of $z^3 - z^2 - z - 2$ is $z = 2$, factor out $(z - 2)$ to obtain $z^3 - z^2 - z - 2 = (z - 2)(z^2 + z + 1)$. The remaining roots are the roots of the simpler polynomial $z^2 + z + 1$. (If you're good at **polynomial division**, or the shorter form called **synthetic division**, that may speed up the factoring. See the author's other book, Intermediate Algebra Skills Practice Workbook.)
- If there isn't any constant term (that is, $a_0 = 0$), one factor is $z = 0$. Reduce all exponents by one. For example, $z^3 + 2z^2 - 3z$ doesn't have a constant term, so it factors as $z(z^2 + 2z - 3)$. One root is $z = 0$. The other roots are the roots of $z^2 + 2z - 3$.
- If $a_n, a_{n-1}, \cdots, a_0$ add up to zero, one root is $z = 1$. For example, $z^6 + 5z^4 - 2z^3 - 4$ has a root of $z = 1$ because $1 + 3 - 2 - 4 = 6 - 6 = 0$.

---

[1] For example, $\frac{3}{4}$ is a reduced fraction because 3 and 4 don't share any common factors. As a counterexample, $\frac{10}{15}$ isn't a reduced fraction; divide 10 and 15 each by 5 to reduce it to $\frac{2}{3}$.

- The root $z = -1$ is a little trickier since odd powers of -1 are negative while even powers are positive. As an example, $z^4 + 4z^3 + 2z^2 - 4z - 3$ has a root of -1 since $(-1)^4 + 4(-1)^3 + 2(-1)^2 - 4(-1) - 3 = 1 - 4 + 2 + 4 - 3 = 0$.
- If there are any fractions, multiply both sides of $p(z) = 0$ by the least common denominator to **remove the fractions**. For example, multiply by 10 on both sides of $\frac{z^2}{2} - \frac{3z}{5} - \frac{4}{5} = 0$ to get the equation $5z^2 - 6z - 8 = 0$, which has only **integer** coefficients. You can verify that $z = 2$ is a root of $5z^2 - 6z - 8 = 0$ and is also a root of $\frac{z^2}{2} - \frac{3z}{5} - \frac{4}{5} = 0$.
- If $a_n, a_{n-1}, \cdots, a_0$ are all pure **integers**,[2] any **integer** roots will evenly divide into a_0. For example, $4z^3 - 8z^2 - 15z + 9$ has a constant term of $a_0 = 9$. The only possible integer roots are $\pm 1, \pm 3$, and ± 9 (since $9 \div 1 = 9, 9 \div 3 = 3$, and $9 \div 9 = 1$). You can check that $z = 3$ is a root: $4(3)^3 + 8(3)^2 - 15(3) + 9 = 4(27) - 8(9) - 45 + 9 = 108 - 72 - 36 = 0$.
- If $a_n, a_{n-1}, \cdots, a_0$ are all pure **integers**, if any roots are reduced **fractions** of the form $\frac{N}{D}$, then a_n will be evenly divisible by D and a_0 will be evenly divisible by N. For example, $2z^3 + 5z^2 - 8z - 6$ has $n = 3, a_3 = 2$, and $a_0 = -6$. A reduced fraction that is a root must have a denominator of $D = 2$ (since the only numbers that evenly divide into $a_3 = 2$ are 1 or 2; you don't need to consider 1 for the denominator because then the fraction would be an integer[3]) and a numerator of $N = 1, N = 2, N = 3$, or $N = 6$ (since the only numbers that evenly divide into $a_0 = -6$ are 1, 2, 3, and 6). The possible fractional roots are $\pm \frac{1}{2}$ or $\pm \frac{3}{2}$ (whereas $\frac{2}{2} = 1$ and $\frac{6}{2} = 3$ would be integer roots). You can check that $z = \frac{3}{2}$ is a root: $2\left(\frac{3}{2}\right)^3 + 5\left(\frac{3}{2}\right)^2 - 8\left(\frac{3}{2}\right) - 6 = 2\left(\frac{27}{8}\right) + 5\left(\frac{9}{4}\right) - 12 - 6 = \frac{27}{4} + \frac{45}{4} - 18 = \frac{72}{4} - 18 = 18 - 18 = 0$.
- If $a_n, a_{n-1}, \cdots, a_0$ all have the same sign, any real roots must be **negative**. For example, if $z^5 + z^2 + 1$ has any real roots, they must be negative numbers.

---

[2] As a counterexample, $4i$ is imaginary so it's **not** an integer (it's not even a real number).
[3] For example, $\frac{2}{1} = 2$. But, in general, you **do** need to consider 1 as a possible numerator.

Chapter 15 – Roots of Polynomials

- After looking for integer or fractional solutions, look for solutions that include **square roots** (like $5 - \sqrt{2}$) or are **complex** (like $1 + 4i$).
- If there are only even powers and all terms have the same sign, all of the roots will be **complex** (or purely imaginary). For example, $z^4 + z^2 + 1$ has no real roots because every term is positive if z is real.
- If $a_n, a_{n-1}, \cdots, a_0$ are all rational (and thus real), if any answers include **square roots**, they will come in pairs along with their **conjugates**. For example, if one root is $8 + \sqrt{3}$, another root is $8 - \sqrt{3}$.
- If $a_n, a_{n-1}, \cdots, a_0$ are all real, if any answers are complex, its complex conjugate will also be a root. For example, if one root is $1 - i$, another root is $1 + i$.
- If any coefficient is **imaginary** or **complex**, expect at least one root to be a complex number. In this case, its complex conjugate isn't necessarily a root.
- If any coefficient is irrational, expect at least one root to be irrational. In this case, its conjugate isn't necessarily a root. If you see a square root like $\sqrt{3}$, this might show up in the roots somewhere, such as $2 - \sqrt{3}$.
- For quadratics or cubics with real coefficients, the discriminant rules will help you determine the number of roots that are real or complex. For quadratics, see Chapter 9. For cubics, see the Appendix.
- Sometimes a polynomial is easier to solve after making a substitution. As an example, $z^4 - 9z^2 + 8 = 0$ is easier to solve if you make the substitution $u = z^2$. With this substitution, the equation becomes $u^2 - 9u + 8 = 0$, which is a quadratic equation. Solve this quadratic to find the possible values of u, and then use $z = \pm\sqrt{u}$ to find the corresponding values for z. See Example 5.
- If you find two numbers a and b such that $p(a)$ and $p(b)$ have **opposite signs**, a root of the polynomial lies **between** a and b. By evaluating $p(z)$ at a variety of values (especially, easy ones likes 0, 1, and -1), this can help you find the roots faster. (Some numerical techniques for root-finding apply this idea.) For example, if $p(2) = -1$ and $p(3) = \frac{1}{6}$, one root lies in the range $2 < z < 3$.
- Sketching a **graph** of the polynomial may help you find the roots faster (as long as the roots are real). There is a root wherever $p(z)$ crosses the z-axis.

Complex Numbers Essentials Math Workbook with Answers

Example 1. Find all of the roots of $p(z) = 6z + 9$.

Solution: Set $p(z)$ equal to zero and solve for z. This gives $6z + 9 = 0$. Subtract 9 from both sides to get $6z = -9$. Divide by 6 on both sides to get $z = -\frac{9}{6} = -\frac{3}{2}$.

Check the answer: $p\left(-\frac{3}{2}\right) = 6\left(-\frac{3}{2}\right) + 9 = -\frac{18}{2} + 9 = -9 + 9 = 0$. ✓

Example 2. Find all of the roots of $p(z) = z^3 - 10z^2 + 25z$.

Solution: Since there isn't a constant term (that is, every term includes z), one root is $z = 0$. Factor out z to get $z(z^2 - 10z + 25)$. The remaining roots are the roots of the quadratic $z^2 - 10z + 25$. This quadratic factors as $(z - 5)(z - 5)$. (Alternatively, you may use the quadratic formula.) There is a **double root** at $z = 5$.

Check the answers: $p(0) = 0^3 - 10(0)^2 + 25(0) = 0 + 0 + 0 = 0$. ✓
$p(5) = 5^3 - 10(5)^2 + 25(5) = 125 - 10(25) + 125 = 250 - 250 = 0$. ✓

Example 3. Find all of the roots of $p(z) = 5z^3 + 33z^2 - 54z + 16$.

Solution: Observe that $5 + 33 - 54 + 16 = 54 - 54 = 0$. This tells us that one of the roots is $z = 1$. If there are any other integer roots, they will evenly divide into $a_0 = 16$. The possibilities include $\pm 1, \pm 2, \pm 4, \pm 8$, or ± 16. If you plug each of these values into $p(z)$ on your calculator (for 4, 8, and 16, it should be easy to see that only negative values have a chance), you should find that $p(-8) = 0$, such that $z = -8$ is another root. Since the degree of the polynomial is 3, there is one more root (unless one of the first two roots happens to be a double root). If the third root is a fraction, its denominator must be 5 (since $a_3 = 5$ is only evenly divisible by 1 and 5; if the denominator were 1, it would be an integer, not a fraction) and its numerator must be 1, 2, 4, 8, or 16 (since $a_0 = 16$ is evenly divisible by these numbers). If you proceed to evaluate plug $\pm\frac{1}{5}, \pm\frac{2}{5}, \pm\frac{4}{5}, \pm\frac{8}{5}$, and $\pm\frac{16}{5}$ into $p(z)$ on your calculator, you should stop when you find that $p\left(\frac{2}{5}\right) = 0$, showing that the last possible root is $z = \frac{2}{5}$.

Check the answers: $p(1) = 5(1)^3 + 33(1)^2 - 54(1) + 16 = 5 + 33 - 54 + 16 = 0$. ✓
$p(-8) = 5(-8)^3 + 33(-8)^2 - 54(-8) + 16 = -2560 + 2112 + 432 + 16 = 0$. ✓
$p\left(\frac{2}{5}\right) = 5\left(\frac{2}{5}\right)^3 + 33\left(\frac{2}{5}\right)^2 - 54\left(\frac{2}{5}\right) + 16 = 0.32 + 5.28 - 21.6 + 16 = 0$. ✓

Chapter 15 – Roots of Polynomials

Example 4. Find all of the roots of $p(z) = 2z^4 - 15z^3 + 25z^2 + 25z - 17$.

Solution: Notice that 2 and 15 make 17, and there are two 25's. There is potential cancellation here. If you try $z = -1$, you should get $2 + 15 + 25 - 25 - 17 = 0$, such that one root is $z = -1$. The only other possible integers are 1 and ± 17 (since $a_0 = -17$), but you can check that those don't work. If there is a fractional root, it must have a denominator of 2 (since $a_4 = 2$ is only divisible by 1 and 2; if the denominator were 1, it would be an integer, not a fraction) and its numerator must be 1 or 17 (since $a_0 = 17$ is prime). If you plug $\pm \frac{1}{2}$ and $\pm \frac{17}{2}$ into $p(z)$ on your calculator, you should find that another root is $z = \frac{1}{2}$. The remaining roots are complex and/or irrational (unless both of the first two roots turn out to be double roots). Let's factor $(z + 1)$ out of $p(z)$. It may take a little trial and error (but if you spend an hour or so learning and practicing polynomial division,[4] this can be done quickly) to get $p(z) = (z + 1)(2z^3 - 17z^2 + 42z - 17)$. Now factor $(2z - 1)$ out of the new polynomial to get $p(z) = (z + 1)(2z - 1)(z^2 - 8z + 17)$. (You might want to try the TOP SECRET method mentioned in footnote 4.[5]) Now we need to find the roots of the quadratic. If you can factor it, great. If you need to use the quadratic formula, that's fine, too.

$$z = \frac{-(-8) \pm \sqrt{(-8)^2 - 4(1)(17)}}{2(1)} = \frac{8 \pm \sqrt{64 - 68}}{2} = \frac{8 \pm \sqrt{-4}}{2} = \frac{8 \pm 2i}{2} = 4 \pm i$$

Check the answers: $p(-1) = 2(-1)^4 - 15(-1)^3 + 25(-1)^2 + 25(-1) - 17$
$= 2(1) - 15(-1) + 25(1) - 25 - 17 = 2 + 15 + 25 - 25 - 17 = 0.$ ✓

$p\left(\frac{1}{2}\right) = 2\left(\frac{1}{2}\right)^4 - 15\left(\frac{1}{2}\right)^3 + 25\left(\frac{1}{2}\right)^2 + 25\left(\frac{1}{2}\right) - 17 = 0.125 - 1.875 + 6.25 + 12.5 - 17 = 0.$ ✓

$(4 \pm i)^2 = 16 \pm 8i - 1 = 15 \pm 8i$

$(4 \pm i)^3 = (4 \pm i)(4 \pm i)^2 = (4 \pm i)(15 \pm 8i) = 60 \pm 32i \pm 15i - 8 = 52 \pm 47i$

---

[4] If you prefer not to divide polynomials, you can multiply instead. The factorization should have the form $2z^4 - 15z^3 + 25z^2 + 25z - 17 = (z + 1)(az^3 + bz^2 + cz + d)$. Multiply this out to get $az^4 + (a + b)z^3 + (b + c)z^2 + (c + d)z + d$. This makes it easy to see that $a = 2$ and $d = -17$, and you can quickly calculate that $b = -17$ and $c = 42$. Shhhh! This method is TOP SECRET.

[5] The Footnote 4 method gives $2z^3 - 17z^2 + 42z - 17 = (2z - 1)(ez^2 + fz + h)$. Multiply this out: $2ez^3 + (2f - e)z^2 + (2h - f)z - h$. Identify $e = 1$ and $h = 17$. This gives $2f - e = -17$ and $2h - f = 42$. Either equation leads to $f = -8$.

$(4 \pm i)^4 = (4 \pm i)^2(4 \pm i)^2 = (15 \pm 8i)(15 \pm 8i) = 225 \pm 240i - 64 = 161 \pm 240i$
$p(4 \pm i) = 2(4 \pm i)^4 - 15(4 \pm i)^3 + 25(4 \pm i)^2 + 25(4 \pm i) - 17$
$= 2(161 \pm 240i) - 15(52 \pm 47i) + 25(15 \pm 8i) + 100 \pm 25i - 17$
$= 322 \pm 480i - 780 \mp 705i + 375 \pm 200i + 83 \pm 25i$
$= 322 + 375 + 83 - 780 \pm 480i \pm 200i \pm 25i \mp 705i$
$= 780 - 780 \pm 705i \mp 705i = 0 \pm 0 = 0.$ ✓

(The sign of $705i$ is opposite to the other imaginary terms.)

Example 5. Find all of the roots of $p(z) = z^4 - 9z^2 + 8$.
Solution: Let $u = z^2$ to write this as $u^2 - 9u + 8$. This factors as $(u - 1)(u - 8)$, such that $u = 1$ or $u = 8$. If we solve $u = z^2$ for z, we get $z = \pm\sqrt{u}$. The $\pm$ is needed to get all 4 roots for z. Note that $\left(-\sqrt{u}\right)^2 = u$. The 4 roots are $z = \pm\sqrt{1} = \pm 1$ and $z = \pm\sqrt{8} =$
$= \pm\sqrt{(4)(2)} = \pm\sqrt{4}\sqrt{2} = \pm 2\sqrt{2}$.
Check the answers: $p(\pm 1) = (\pm 1)^4 - 9(\pm 1)^2 + 8 = 1 - 9 + 8 = 0.$ ✓
$p(\pm 2\sqrt{2}) = (\pm 2\sqrt{2})^4 - 9(\pm 2\sqrt{2})^2 + 8 = (\pm 2)^4(\sqrt{2})^4 - 9(\pm 2)^2(\sqrt{2})^2 + 8$
$= 16\sqrt{2}\sqrt{2}\sqrt{2}\sqrt{2} - 9(4)(2) + 8 = 16(2)(2) - 72 + 8 = 64 - 72 + 8 = 0.$ ✓

Chapter 15 Problems

Directions: Find all of the roots of each polynomial.

(1) $p(z) = 6z - 42$

(2) $p(z) = z^3 - 18z^2 + 101z - 180$

(3) $p(z) = 12z^4 - 67z^3 - 32z^2 + 12z$

(4) $p(z) = 16z^3 - 56z^2 + 65z - 25$

(5) $p(z) = 6z^5 - 5z^4 - 78z^3 + 65z^2 + 216z - 180$

(6) $p(z) = z^3 + z^2 - 3z - 3$

(7) $p(z) = 2z^3 + 3z^2 + 8z + 12$

(8) $p(z) = z^6 - 18z^4 + 29z^2 + 48$

(9) $p(z) = 6z^3 - 49z^2 + 158z - 25$

(10) $p(z) = 16z^4 + 8(5 - 4i)z^3 + (13 - 80i)z^2 - 2(3 + 13i)z + 12i$

16 Properties of Complex Numbers

The properties of complex numbers listed in this chapter are in the form of handy formulas or identities. Some of these will be proven in the examples or exercises.

Integer powers of i are:

$$i^n = \begin{cases} 1 & if\ n = 0, 4, 8, 12, 16, \cdots \quad n \div 4 \text{ has a remainder of } 0 \\ i & if\ n = 1, 5, 9, 13, 17, \cdots \quad n \div 4 \text{ has a remainder of } 1 \\ -1 & if\ n = 2, 6, 10, 14, 18, \cdots \quad n \div 4 \text{ has a remainder of } 2 \\ -i & if\ n = 3, 7, 11, 15, 19, \cdots \quad n \div 4 \text{ has a remainder of } 3 \end{cases}$$

A complex number z may be expressed in **rectangular**, **polar**, or **exponential** form.

$$z = x + iy = |z| \cos \theta + i|z| \sin \theta = |z|e^{i\theta}$$

The real and imaginary parts are related to the **modulus** and **argument** by:

$$x = |z| \cos \theta \quad , \quad y = |z| \sin \theta \quad , \quad |z| = \sqrt{x^2 + y^2} \quad , \quad \theta = \tan^{-1}\left(\frac{y}{x}\right)$$

Let $z_1 = x_1 + iy_1$ and $z_2 = x_2 + iy_2$ be two different complex numbers with moduli $|z_1|$ and $|z_2|$ and arguments θ_1 and θ_2. To add or subtract two complex numbers,

$$z_1 \pm z_2 = x_1 \pm x_2 + i(y_1 \pm y_2)$$
$$z_1 \pm z_2 = |z_1| \cos \theta_1 \pm |z_2| \cos \theta_2 + i(|z_1| \sin \theta \pm |z_2| \sin \theta_2)$$
$$z_1 \pm z_2 = |z_1|e^{i\theta_1} \pm |z_2|e^{i\theta_2}$$

To multiply two complex numbers,

$$z_1 z_2 = x_1 x_2 - y_1 y_2 + i(x_1 y_2 + y_1 x_2)$$
$$z_1 z_2 = |z_1||z_2|[\cos(\theta_1 + \theta_2) + i \sin(\theta_1 + \theta_2)]$$
$$z_1 z_2 = |z_1||z_2|e^{i(\theta_1 + \theta_2)}$$

To divide two complex numbers,

$$\frac{z_1}{z_2} = \frac{x_1 x_2 + y_1 y_2 + i(x_2 y_1 - x_1 y_2)}{x_2^2 + y_2^2}$$
$$\frac{z_1}{z_2} = \frac{|z_1|}{|z_2|}[\cos(\theta_1 - \theta_2) + i \sin(\theta_1 - \theta_2)]$$
$$\frac{z_1}{z_2} = \frac{|z_1|}{|z_2|}e^{i(\theta_1 - \theta_2)}$$

Chapter 16 – Properties of Complex Numbers

If x and y are each real, the **complex conjugate** is:
$$\bar{z} = x - iy = |z|\cos\theta - |z|i\sin\theta = |z|e^{-i\theta}$$
Any number times its complex conjugate is a real nonnegative number.
$$z\bar{z} = |z|^2 = x^2 + y^2 \geq 0$$
Identities involving complex conjugates include:
$$|z| = \sqrt{z\bar{z}} \quad , \quad \bar{z} = \frac{|z|^2}{z} \quad , \quad \bar{\bar{z}} = z \quad , \quad |\bar{z}^2| = |\bar{z}|^2$$
$$\overline{z_1 \pm z_2} = \bar{z_1} \pm \bar{z_2} \quad , \quad \overline{z_1 z_2} = \bar{z_1}\,\bar{z_2} \quad , \quad \overline{z_1/z_2} = \bar{z_1}/\bar{z_2}$$

If x and y are each real, the real and imaginary parts of a complex number are:
$$\text{Re}(z) = x = |z|\cos\theta \quad , \quad \text{Im}(z) = y = |z|\sin\theta$$
$$\text{Re}(z) = \text{Re}(\bar{z}) = \frac{z + \bar{z}}{2} \quad , \quad \text{Im}(z) = -\text{Im}(\bar{z}) = \frac{z - \bar{z}}{2i}$$

Identities involving the **modulus** include:
$$|\bar{z}| = |z| \quad , \quad |z^2| = |z|^2$$
$$|z_1 \pm z_2| = \sqrt{(x_1 \pm x_2)^2 + (y_1 \pm y_2)^2}$$
$$|z_1 \pm z_2| = \sqrt{(|z_1|\cos\theta_1 \pm |z_2|\cos\theta_2)^2 + (|z_1|\sin\theta \pm |z_2|\sin\theta_2)^2}$$
$$|z_1 z_2| = |z_1||z_2|$$
$$\left|\frac{z_1}{z_2}\right| = \frac{|z_1|}{|z_2|}$$

Euler's formula is:
$$e^{i\theta} = \cos\theta + i\sin\theta$$
De Moivre's theorem is:
$$(\cos\theta + i\sin\theta)^n = \cos(n\theta) + i\sin(n\theta)$$
$$z^n = |z|^n e^{in\theta} = |z|^n(\cos\theta + i\sin\theta)^n = |z|^n \cos(n\theta) + i|z|^n \sin(n\theta)$$

Identities relating **trig** and **hyperbolic** functions to **exponentials** include:
$$\cos\theta = \frac{e^{i\theta} + e^{-i\theta}}{2} \quad , \quad \sin\theta = \frac{e^{i\theta} - e^{-i\theta}}{2i} \quad , \quad \tan\theta = \frac{1}{i}\frac{e^{i\theta} - e^{-i\theta}}{e^{i\theta} + e^{-i\theta}}$$
$$\cosh\theta = \frac{e^\theta + e^{-\theta}}{2} \quad , \quad \sinh\theta = \frac{e^\theta - e^{-\theta}}{2} \quad , \quad \tanh\theta = \frac{e^\theta - e^{-\theta}}{e^\theta + e^{-\theta}}$$

Identities relating **trig** functions to **hyperbolic** functions include (see Chapter 13):
$$\cos(i\theta) = \cosh\theta \quad , \quad \sin(i\theta) = i\sinh\theta$$
$$\cosh(i\theta) = \cos\theta \quad , \quad \sinh(i\theta) = i\sin\theta$$
Angle sum identities (derived in examples or exercises in Chapter 13) include:
$$\cos(\theta + i\varphi) = \cos\theta\cosh\varphi - i\sin\theta\sinh\varphi$$
$$\sin(\theta + i\varphi) = \sin\theta\cosh\varphi + i\cos\theta\sinh\varphi$$
$$\cosh(\theta + i\varphi) = \cosh\theta\cos\varphi + i\sinh\theta\sin\varphi$$
$$\sinh(\theta + i\varphi) = \sinh\theta\cos\varphi + i\cosh\theta\sin\varphi$$
Negative angle identities include:
$$\cos(-\theta) = \cos\theta \quad , \quad \sin(-\theta) = -\sin\theta$$
$$\cosh(-\theta) = \cosh\theta \quad , \quad \sinh(-\theta) = -\sinh\theta$$

In order for two complex numbers to be equal, $z_1 = z_2$, then $x_1 = x_2$ and $y_1 = y_2$ must both be true. According to the **transitive** property, if $z_1 = z_2$ and $z_2 = z_3$, then $z_1 = z_3$. According the **commutative** rules of addition and multiplication, the order in which complex numbers are added or multiplied doesn't matter.
$$z_1 + z_2 = z_2 + z_1 \quad , \quad z_1 z_2 = z_2 z_1$$
According to the **associative** rules of addition and multiplication:
$$(z_1 + z_2) + z_3 = z_1 + (z_2 + z_3) = z_2 + (z_1 + z_3)$$
$$(z_1 z_2) z_3 = z_1 (z_2 z_3) = z_2 (z_1 z_3)$$
The **distributive** property also holds for complex numbers.
$$z_1(z_2 + z_3) = z_1 z_2 + z_1 z_3 = (z_2 + z_3) z_1$$

The **additive** inverse is the negative of a complex number. The **multiplicative** inverse is the reciprocal of a complex number, which involves division.
$$z + (-z) = 0 \quad , \quad -z = -x - iy = -|z|\cos\theta - i|z|\sin\theta = |z|e^{i(\theta + \pi)}$$
$$zz^{-1} = 1 \quad , \quad z^{-1} = \frac{1}{z} = \frac{x - iy}{x^2 + y^2} = \frac{\cos\theta - i\sin\theta}{|z|} = \frac{e^{-i\theta}}{|z|}$$

Like the trig functions, the exponential form of a complex number is **periodic**; adding $2\pi i k$, where k is an integer, to the argument has no effect. Another way to say this is to say that the exponential function e^w (where w may be complex) has a period of $2\pi i$.
$$e^{i(\theta + 2\pi k)} = e^{i\theta}$$

In the context of complex numbers, the **logarithm** reflects this periodicity; as a result, it is **multi-valued**. When taking the logarithm of a complex number (or when taking the logarithm of a negative real number), use the formula below.

$$\ln z = \ln(|z|e^{i\theta}) = i\theta + 2\pi i k + \ln|z|$$

Complex exponentiation is similarly **multi-valued**. Note that $e^{w \ln z} = e^{\ln z^w} = z^w$ since $w \ln z = \ln z^w$ and $e^{\ln u} = u$. For a **complex exponent** w, use the complex logarithm:

$$z^w = e^{w \ln z} = e^{w(i\theta + 2\pi i k + \ln|z|)}$$

The n **roots** of a (nonzero) number are (where $k = 0, 1, 2, 3, \cdots, n-1$):

$$\sqrt[n]{z} = z^{1/n} = \sqrt[n]{|z|} \left[\cos\left(\frac{\theta + 2\pi k}{n}\right) + i \sin\left(\frac{\theta + 2\pi k}{n}\right) \right]$$

Beware: Many of the identities relating to logarithms and exponents that hold for real numbers involve multiple roots or periodicity in the context of complex numbers. For example, if x, y, a, and b are real, $\ln(x^a) = a \ln x$, $e^{\ln x} = x$, $(x^a)^b = x^{ab}$, and $(xy)^a = x^a y^a$, but in the context of complex numbers, **such relations may not work as usual**.

First, consider $(tz)^w = t^w z^w$ with $t = -4$, $z = -16$, and $w = \frac{1}{2}$. At first glance, it may seem like these are all real numbers, but upon closer inspection, $t^w = (-4)^{1/2} = \sqrt{-4} = 2i$ and $z^w = (-16)^{1/2} = \sqrt{-16} = 4i$ are imaginary numbers. It appears that the right-hand side equals $t^w z^w = (2i)(4i) = 8i^2 = -8$, whereas the left-hand side appears to be $(tz)^w = [(-4)(-16)]^{1/2} = 64^{1/2} = \sqrt{64} = 8$. Obviously, 8 doesn't equal -8. Where did we go wrong? You may recall that we encountered a similar situation at the end of Chapter 1. In this case, roots are multi-valued. A square root has two answers; one is positive and the other is negative. The usual convention of writing $\sqrt{64}$ to mean just the positive root can be exploited to create seeming paradoxes in the context of complex numbers. What we have really shown is that $\pm 8 = \pm 8$.

As a second example, we will consider a seeming paradox discovered in 1827.[1] Start with $e^{1+2\pi i}$. If we apply the ordinary relation $e^a e^z = e^{a+z}$ that works for real numbers, we

---

[1] Steiner, J.; Clausen, T.; Abel, Niels Henrik, "Aufgaben und Lehrsätze, erstere aufzulösen, letztere zu beweisen," Journal für die reine und angewandte Mathematik 2: 286-287 (1827).

get $e^{1+2\pi i} = e^1 e^{2\pi i} = (e)(\cos 2\pi + i \sin 2\pi) = e(1) = e$ (using Euler's formula to see that $e^{2\pi i} = 1$). Raise both sides to the power of $1 + 2\pi i$. The right-hand side is again e.
$$\left(e^{1+2\pi i}\right)^{1+2\pi i} = e^{1+2\pi i} = e(1) = e$$
If we apply the relation $(e^w)^z = e^{wz}$ that works for real numbers, we get
$$\left(e^{1+2\pi i}\right)^{1+2\pi i} = e^{(1+2\pi i)(1+2\pi i)} = e^{1+4\pi i + 4\pi^2 i^2} = e^{1+4\pi i - 4\pi^2} = e$$
Now apply the relation $e^a e^z e^b = e^{a+z+b}$.
$$e^1 e^{4\pi i} e^{-4\pi^2} = e(\cos 4\pi + i \sin 4\pi)e^{-4\pi^2} = e(1)e^{-4\pi^2} = ee^{-4\pi^2} = e$$
Divide by e on both sides of $ee^{-4\pi^2} = e$ to get $e^{-4\pi^2} = \frac{1}{e^{4\pi^2}} = 1$. Here's the problem. If you get out your calculator, you will see that $e^{-4\pi^2} \approx 7.157 \times 10^{-18}$ is an extremely tiny number. It most certainly doesn't equal one. So where does this proof go wrong?

- First, note that changing the powers from $1 + 2\pi i$ to $1 + 2\pi i k$ on the left-hand side won't by itself alleviate the problem. This will just give us a factor of k^2 along with the $-4\pi^2$. We should include this k, but it isn't the only problem.
- We shouldn't proceed to separate $e^{w+z} = e^w e^z$ into factors that effectively remove the full periodicity in selected locations. The proof basically begins by writing $e^{1+2\pi i}$ on the left, but simply calling it e^1 on the right; including the periodicity on one side only and then manipulating that one side while leaving the other unchanged. What we really have is $e^{1+2\pi i k} = e^{1+2\pi i k}$, and if you want to raise both sides to powers, it is $\left(e^{1+2\pi i k}\right)^{1+2\pi i k} = \left(e^{1+2\pi i k}\right)^{1+2\pi i k}$. Writing $\left(e^{1+2\pi i k}\right)^{1+2\pi i k} = e$ includes the full periodicity on one side only. (It really is as narrow as not including the $\pm$ on both sides of the previous example with the square roots.)

The main idea is that complex exponents and logarithms are multi-valued in nature. This seeming 'proof' exploits the periodicity of complex numbers.

Often, the argument of a complex number is restricted to its **principal value**, for which $-\pi < \theta \leq \pi$. However, it's important to note that **this restriction doesn't avoid the problem of paradoxes**. For example, see Exercise 18, where $\theta = \pi$ lies in the principal value range. To resolve paradoxes, it is necessary to work with the **full periodicity** of complex functions (rather than restrict them) and to consider **all possible roots**.

Chapter 16 – Properties of Complex Numbers

Example 1. Evaluate $\ln(-1)$.

Solution: How does this relate to complex numbers? The natural log of a negative number isn't real. If you try it on a standard calculator, you will get a domain error. Use the (periodic) formula for a complex logarithm. Note that $|z| = 1$ and $\theta = \pi$ is the principal[2] argument for the number -1 (which lies on the negative x-axis). Another way to see that $\theta = \pi$ is via Euler's formula: $e^{i\pi} = \cos \pi + i \sin \pi = -1$.

$$\ln z = \ln(|z|e^{i\theta}) = i\theta + 2\pi i k + \ln|z|$$

$$\ln(-1) = \ln(1e^{i\pi}) = i\pi + 2\pi i k + \ln 1 = i\pi + 2\pi i k + 0 = i\pi + 2\pi i k = i\pi(1 + 2k)$$

Example 2. Evaluate 1^i.

Solution: Use the formula for complex exponentiation. Even though $\ln 1$ ordinarily just equals zero in the context of real numbers, in the context of complex numbers we need to incorporate the periodicity inherent in complex exponentiation. With $z = 1$, the principal argument is $\theta = 0$. (As usual, $\theta + 2\pi k$ accounts for all possibilities.)

$$z^w = e^{w \ln z} = e^{w(i\theta + 2\pi i k + \ln|z|)}$$

$$1^i = e^{i \ln 1} = e^{i(0i + 2\pi i k + \ln 1)} = e^{i(2\pi i k + 0)} = e^{-2\pi k}$$

Example 3. Show that $z_1 z_2 = x_1 x_2 - y_1 y_2 + i(x_1 y_2 + y_1 x_2)$.

Solution: Plug in $z_1 = x_1 + iy_1$ and $z_2 = x_2 + iy_2$.

$$z_1 z_2 = (x_1 + iy_1)(x_2 + iy_2) = x_1 x_2 + ix_1 y_2 + iy_1 x_2 + i^2 y_1 y_2$$

$$z_1 z_2 = x_1 x_2 + ix_1 y_2 + iy_1 x_2 - y_1 y_2 = x_1 x_2 - y_1 y_2 + i(x_1 y_2 + y_1 x_2)$$

Example 4. Show that $|z_1 + z_2| = \sqrt{(|z_1| \cos \theta_1 + |z_2| \cos \theta_2)^2 + (|z_1| \sin \theta + |z_2| \sin \theta_2)^2}$.

Solution: First use $z_1 = x_1 + iy_1$ and $z_2 = x_2 + iy_2$ to find $z_1 + z_2$.

$$z_1 + z_2 = x_1 + iy_1 + x_2 + iy_2 = (x_1 + x_2) + i(y_1 + y_2)$$

To find the modulus of a complex number, square the real part, square the imaginary part, add these squares together, and take the square root.

$$|z_1 + z_2| = \sqrt{(x_1 + x_2)^2 + (y_1 + y_2)^2}$$

---

[2] The $2\pi i k$ incorporates the periodicity inherent in complex exponentiation and logarithms. This says that you can add or subtract any integer multiple of 2π to the principal argument and get an equivalent angle.

Now convert the Cartesian components to polar form using $x_1 = |z_1|\cos\theta_1$, $x_2 = |z_2|\cos\theta_2$, $y_1 = |z_1|\sin\theta_1$, and $y_2 = |z_2|\sin\theta_2$.
$$|z_1 + z_2| = \sqrt{(|z_1|\cos\theta_1 + |z_2|\cos\theta_2)^2 + (|z_1|\sin\theta + |z_2|\sin\theta_2)^2}$$

Example 5. Show that $|z^2| = |z|^2 = |\bar{z}|^2 = |\bar{z}^2|$.

Solution: Begin with $z = x + iy$. Its modulus square is
$$|z|^2 = x^2 + y^2$$
The complex conjugate is $\bar{z} = x - iy$. The modulus square of the complex conjugate is
$$|\bar{z}|^2 = x^2 + (-y)^2 = x^2 + y^2$$
The square of z is
$$z^2 = zz = (x+iy)(x+iy) = x^2 + 2ixy - y^2 = (x^2 - y^2) + 2ixy$$
The modulus of the square of z is
$$|z^2| = \sqrt{(x^2-y^2)^2 + (2xy)^2} = \sqrt{x^4 - 2x^2y^2 + y^4 + 4x^2y^2} = \sqrt{x^4 + 2x^2y^2 + y^4}$$
$$|z^2| = \sqrt{(x^2+y^2)^2} = x^2 + y^2$$
Finally, the square of the complex conjugate is
$$(\bar{z})^2 = (x-iy)(x-iy) = x^2 - 2ixy - y^2 = (x^2 - y^2) - 2ixy$$
The modulus of the square of the complex conjugate is
$$|\bar{z}^2| = \sqrt{(x^2-y^2)^2 + (-2xy)^2} = \sqrt{x^4 - 2x^2y^2 + y^4 + 4x^2y^2} = \sqrt{x^4 + 2x^2y^2 + y^4}$$
$$|\bar{z}^2| = \sqrt{(x^2+y^2)^2} = x^2 + y^2$$

Example 6. Resolve the apparent paradox in the work below.
- $i\ln(e^{2\pi i}) = i\ln(\cos 2\pi + i\sin 2\pi) = i\ln 1 = i(0) = 0$ using Euler's formula to write $e^{2\pi i} = \cos 2\pi + i\sin 2\pi = 1$ and the fact that $\ln 1 = 0$.
- $i\ln(e^{2\pi i}) = i(2\pi i)\ln e = -2\pi(1) = -2\pi$ using $\ln(a^z) = z\ln a$ to get $\ln(e^{2\pi i}) = 2\pi i \ln e$ and the fact that $\ln e = 1$.
- But 0 doesn't equal 1. Where did we go wrong?

Solution: Each bullet point ignores the full periodicity of complex exponentiation and logarithms. Although $e^{2\pi i}$ equals 1 (through Euler's formula) and $\ln 1 = 0$, since $e^{2\pi i}$ is written in the context of complex numbers, we should use the complex form of the logarithm, $\ln z = \ln(|z|e^{i\theta}) = i\theta + 2\pi i k + \ln|z|$. Compare the general form $\ln(|z|e^{i\theta})$ to $\ln(e^{2\pi i})$ to see that $|z| = 1$ and $\theta = 2\pi$. Both bullet points should be:

$$\ln(e^{2\pi i}) = 2\pi i + 2\pi i k + \ln 1 = 2\pi i + 2\pi i k + 0 = 2\pi i(k+1)$$

Let $k' = k + 1$ to write this more concisely as $\ln(e^{2\pi i}) = 2\pi i k'$. (Since k and k' each cover the full set of integers, the prime isn't really meaningful here.) Now multiply by i to get $i\ln(e^{2\pi i}) = i(2\pi i k') = 2\pi i^2 k' = -2\pi k'$. For the special case $k' = 0$, we get the first bullet point, and for the special case $k' = 1$, we get the second bullet point. The correct answer is $-2\pi k'$ where k' is an integer. This infinite set of possibilities covers every case.

Chapter 16 Problems

Directions: Evaluate each expression.

(1) i^i

(2) $(-1)^i$

(3) $\ln(i)$

(4) $\ln(e^{i\pi})$

(5) $\left(\dfrac{i}{2} - \dfrac{\sqrt{3}}{2}\right)^i$

(6) $\ln\left(\dfrac{\sqrt{3}}{2} + \dfrac{i}{2}\right)$

(7) $\ln\left(\dfrac{\sqrt{2}}{2} - i\dfrac{\sqrt{2}}{2}\right)$

(8) i^{3-2i}

Chapter 16 – Properties of Complex Numbers

Directions: Derive each identity.

(9) $\dfrac{z_1}{z_2} = \dfrac{x_1 x_2 + y_1 y_2 + i(x_2 y_1 - x_1 y_2)}{x_2^2 + y_2^2}$

(10) $z_1 z_2 = |z_1||z_2|[\cos(\theta_1 + \theta_2) + i\sin(\theta_1 + \theta_2)]$

(11) $\bar{z} = \dfrac{|z|^2}{z}$

(12) $\overline{z_1 + z_2} = \overline{z_1} + \overline{z_2}$

(13) $\overline{z_1 z_2} = \overline{z_1}\,\overline{z_2}$

(14) $\text{Im}(z) = \dfrac{z - \bar{z}}{2i}$

(15) $|z_1 z_2| = |z_1||z_2|$

(16) $z^{-1} = \dfrac{1}{z} = \dfrac{x - iy}{x^2 + y^2}$

Chapter 16 – Properties of Complex Numbers

Directions: Each problem presents what seems to be a paradox. Resolve the paradox.

(17) $\dfrac{\sqrt{1}}{\sqrt{-1}} = \dfrac{1}{i} = \dfrac{i}{i^2} = \dfrac{i}{-1} = -i$

but $\dfrac{\sqrt{1}}{\sqrt{-1}} = \sqrt{\dfrac{1}{-1}} = \sqrt{-1} = i$

(18) $e^{i\pi} = -1$ implies that $i\pi = \ln(-1)$

$\ln(-1) + \ln(-1) = i\pi + i\pi = 2i\pi$

but according to $\ln w + \ln z = \ln wz$,

$\ln(-1) + \ln(-1) = \ln[(-1)(-1)] = \ln 1 = 0$

(19) $e^{3i\pi/2} = e^{-i\pi/2} = -i$

$\left(e^{3i\pi/2}\right)^{1/3}$ should $= \left(e^{-i\pi/2}\right)^{1/3}$

Apply $(e^w)^a = e^{aw}$ to each side

$\left(e^{3i\pi/2}\right)^{1/3} = e^{i\pi/2}$ since $\dfrac{3i\pi}{2}\dfrac{1}{3} = \dfrac{i\pi}{2}$

$\left(e^{-i\pi/2}\right)^{1/3} = e^{-i\pi/6}$ since $\dfrac{i\pi}{2}\dfrac{1}{3} = \dfrac{i\pi}{6}$

$e^{i\pi/2}$ should $= e^{-i\pi/6}$

But $e^{i\pi/2} = i$ and $e^{-i\pi/6} = \dfrac{\sqrt{3}}{2} - \dfrac{i}{2}$

(20) $1^x = 1$ for all x

$e^{2i\pi} = \cos 2\pi + i \sin 2\pi = 1 + 0i = 1$

But $\left(e^{2i\pi}\right)^{2i\pi} = e^{(2i\pi)(2i\pi)} = e^{4i^2\pi^2} = e^{-4\pi^2}$

and $e^{-4\pi^2} = \dfrac{1}{e^{4\pi^2}} \neq 1$ contradicts $1^x = 1$

for $x = 2i\pi$

122

17 Applications

In this chapter, we will discuss a variety of applications of complex numbers in the fields of math and physics.

As we saw in Chapter 15, complex numbers are significant in the **roots of polynomials**. Consider for a moment what finding roots of polynomials is like without using any complex numbers. One of the strange things is that every cubic polynomial (with real coefficients) has at least one real root, whereas many quadratics and quartics (with real coefficients) have no real roots. For example, $x^2 + 4$ and $3x^4 + 6x^2 + 1$ don't have any real roots because no real number squared or raised to the fourth power can be negative to cancel the positive constant. When we allow for complex roots, the roots of polynomials bring a feeling of completeness in the following ways. First, every polynomial has either real or complex roots (or both), so no polynomial is left out. Also, a polynomial of degree n will have n roots (either real or complex), though some of the roots may be the same (with multiplicity of 2 or more). In this way, complex roots play an important role in the **fundamental theorem of algebra** (Chapter 15).

Complex numbers are also important in trigonometry. Although we explored some elements of complex trigonometry in Chapter 12, what may seem fascinating is that complex numbers are helpful in the trigonometry of purely **real numbers**. Complex numbers make it easier to derive some standard **trig identities**. Recall the following identities from Chapters 10-12 relating to Euler's formula and de Moivre's theorem:

$$e^{i\theta} = \cos\theta + i\sin\theta \quad , \quad e^{-i\theta} = \cos\theta - i\sin\theta$$

$$\cos\theta = \frac{e^{i\theta} + e^{-i\theta}}{2} \quad , \quad \sin\theta = \frac{e^{i\theta} - e^{-i\theta}}{2i} \quad , \quad \tan\theta = \frac{1}{i}\frac{e^{i\theta} - e^{-i\theta}}{e^{i\theta} + e^{-i\theta}}$$

$$\left(re^{i\theta}\right)^n = r^n[\cos(n\theta) + i\sin(n\theta)]$$

In Chapter 1, we showed how Euler's formula makes it easy to derive the **angle sum formulas** for $\cos(\alpha + \beta)$ and $\sin(\alpha + \beta)$, whereas the geometric method using real numbers is difficult for most students. In Example 1 and in the exercises, we will see more examples of how complex numbers help to derive a variety of trig identities.

Chapter 17 – Applications

In **calculus**,[1] complex numbers help with a class of **improper integrals**. In calculus, an integral over one variable represents the area under the curve between the initial and final points. If the function (called the integrand) for the curve becomes infinite over the interval of integration or if either limit of integration is infinite, the integral is said to be improper. For example, the function $\frac{1}{x^n}$ grows to infinity as x approaches zero if $n > 0$. If the function $\frac{1}{x^n}$ is integrated along an interval that includes $x = 0$ (such as the interval $0 \leq x \leq 1$) or if either endpoint is infinite (as in the interval $1 \leq x < \infty$), the integral is improper. One of the techniques for dealing with improper integrals of real-valued functions is to carry out an integral in the complex plane using a suitable complex function. This is the basic idea behind **contour integrals**. Contour integrals may be performed directly along a curve in the complex plane, or they may be performed by applying the **residue theorem** or **Cauchy's integral formula**.

In **linear algebra** (or **matrix algebra**), some matrices with purely real elements have complex **eigenvalues**. The eigenvalues of a square matrix satisfy the equation $A|x\rangle = \lambda|x\rangle$, where A is the matrix, λ is an eigenvalue, and $|x\rangle$ is a corresponding eigenvector. The eigenvalues can be found by setting the determinant $|A - \lambda I|$ equal to zero, where I is the identity matrix (which has 1's along the main diagonal and 0's elsewhere). For example, for a 2×2 matrix, this determinant is

$$\begin{vmatrix} q - \lambda & r \\ s & t - \lambda \end{vmatrix} = 0$$

$$(q - \lambda)(t - \lambda) - rs = 0$$

$$qt - q\lambda - t\lambda + \lambda^2 - rs = 0$$

$$\lambda^2 - (q + t)\lambda + (qt - rs) = 0$$

The discriminant for this quadratic (recall Chapter 9) is

$$[-(q + t)]^2 - 4(1)(qt - rs) = q^2 + 2qt + t^2 - 4qt + 4rs$$

$$= q^2 - 2qt + t^2 + 4rs = (q - t)^2 + 4rs$$

---

[1] Books on complex analysis generally include contour integrals and the residue theorem; such books expect readers to be fluent in their first few semesters of calculus. The book you're reading now on complex numbers was designed to be accessible for students who are familiar with precalculus (especially, trigonometry and logarithms); contour integrals are beyond the scope of this workbook.

If exactly one of r or s is negative, then the product rs is negative, and if $4rs$ is more negative than $(q - t)^2$ is positive, then the discriminant will be negative. In this case, the eigenvalues of the 2 × 2 matrix are complex even though all four coefficients are real. For example, consider the 2 × 2 matrix below, which has the effect of rotating a vector by 30° in the xy plane.

$$A = \begin{pmatrix} \cos 30° & -\sin 30° \\ \sin 30° & \cos 30° \end{pmatrix} = \begin{pmatrix} \frac{\sqrt{3}}{2} & -\frac{1}{2} \\ \frac{1}{2} & \frac{\sqrt{3}}{2} \end{pmatrix}$$

In this example, $q = t = \frac{\sqrt{3}}{2}$, $r = -\frac{1}{2}$, and $s = \frac{1}{2}$. The quadratic equation is

$$\lambda^2 - (q+t)\lambda + (qt - rs) = \lambda^2 - \left(\frac{\sqrt{3}}{2} + \frac{\sqrt{3}}{2}\right)\lambda + \left[\frac{\sqrt{3}}{2}\frac{\sqrt{3}}{2} - \left(-\frac{1}{2}\right)\left(\frac{1}{2}\right)\right] = 0$$

$$\lambda^2 - \lambda\sqrt{3} + \left(\frac{3}{4} + \frac{1}{4}\right) = \lambda^2 - \lambda\sqrt{3} + 1 = 0$$

The discriminant formula $b^2 - 4ac = \left(-\sqrt{3}\right)^2 - 4(1)(1) = 3 - 4 = -1$ agrees with $(q-t)^2 + 4rs = \left(\frac{\sqrt{3}}{2} - \frac{\sqrt{3}}{2}\right)^2 + 4\left(-\frac{1}{2}\right)\left(\frac{1}{2}\right) = 0^2 - 1 = -1$. Since the discriminant is negative, both eigenvalues are complex, yet all of the elements of the rotation matrix are real and the matrix describes a very real operation (rotating a vector by 30°). The existence of complex eigenvalues has practical applications, such as **matrix exponentials** (for example, these are important in **Lie algebra**).

Matrices with imaginary (or complex) elements can represent physical phenomena. This is common in **quantum mechanics**. For example, the **Pauli spin matrices** include one 2 × 2 matrix with imaginary elements (see σ_2 below):

$$\sigma_1 = \begin{pmatrix} 0 & 1 \\ 1 & 0 \end{pmatrix} \quad , \quad \sigma_2 = \begin{pmatrix} 0 & -i \\ i & 0 \end{pmatrix} \quad , \quad \sigma_3 = \begin{pmatrix} 1 & 0 \\ 0 & -1 \end{pmatrix}$$

The Pauli spin matrices are used when a particle (like an electron or proton) with intrinsic spin angular momentum interacts with an electromagnetic field.

In physics, if a quantity can be directly measured, it is real. Not all quantities can be measured directly. Of those quantities that can only be detected indirectly, some turn

Chapter 17 – Applications

out to be complex. For example, the **wave function** ψ in **quantum mechanics** is often complex. The wave function satisfies **Schrödinger's equation** (which is a second-order differential equation). Although the wave function ψ is often complex, ψ itself can't be measured directly. The wave function ψ can be used to calculate other quantities that can be measured directly, and those quantities always turn out to be real. As an example, the quantity $\psi\overline{\psi}$ (which is the product of the wave function and its complex conjugate) is always a real number (recall Chapters 3 and 16). Since $\psi\overline{\psi}$ represents probability, $\psi\overline{\psi}$ can be measured. Even when ψ is complex, the probability $\psi\overline{\psi}$ is real. Other quantities involving ψ and $\overline{\psi}$ which are measurable also turn out to be real. The mathematical formulation of quantum mechanics is inherently complex; complex numbers appear in **Schrödinger's equation**, **Hilbert space**, the **matrix** representation, and the formulation of momentum and energy **operators**, for example.

Another branch of physics where complex numbers are common is **AC circuits**. In an AC circuit, the current and voltage (or potential difference) alternate. An AC circuit typically includes an AC power supply and circuit elements such as resistors, inductors, or capacitors. The voltage across a particular circuit element may not be in phase with the current; if you plot the voltage and the current as functions of time, if they are not in phase, one will be shifted horizontally compared to the other. Complex numbers are used to represent such **phase shifts**. The voltage is expressed as $V_0 e^{i(\omega t+\varphi)}$, where V_0 is the amplitude of the voltage, ω is the angular frequency (which is 2π times the frequency), t is time, and φ is the phase angle (which indicates to what extent the voltage is out of phase with the current). Euler's formula can be used to write this as $V_0 \cos(\omega t + \varphi) + V_0 i \sin(\omega t + \varphi)$. The real part, which equals $V_0 \cos(\omega t + \varphi)$, can be measured directly. In an AC circuit, the magnitude of the complex voltage equals the magnitude of the complex current times the magnitude of the **impedance**: $|V| = |I||Z|$. The complex impedance[2] is $Z = R + iX$, where R is resistance and X is reactance. The reactance is $X = \omega L - \frac{1}{\omega C}$, where L is the inductance and C is the capacitance. The

---

[2] We're using i for the imaginary number to be consistent with the rest of this workbook. However, texts on AC circuits tend to use j instead, so as not to be easily confused with the current I. (Why don't they use C for current? Because C stands for capacitance, or the unit of charge, which is the Coulomb.)

term ωL is referred to as the inductive reactance (relating to the inductor) while the term $\frac{-1}{\omega C}$ is referred to as the capacitive reactance (relating to the capacitor). It can be shown that the magnitude of the current is maximum when the impedance is purely real; this is the case when the reactance is zero, meaning that $\omega L = \frac{1}{\omega C}$. The frequency for which $\omega L = \frac{1}{\omega C}$ is called the **resonance frequency**; it equals $\omega = \frac{1}{\sqrt{LC}}$.

The **Fourier transform** is widely applied in mathematics, physics, and engineering. A Fourier transform involves calculus; specifically, a Fourier transform is an integral over a real function of one variable (usually, time) that produces a complex function of another variable (usually, frequency). Complex numbers are an intrinsic part of Fourier transforms since one of the factors has the form $e^{i\omega t}$, where ω is the angular frequency and t is time. Fourier transforms are used in **signal processing**, in **quantum mechanics**, and to analyze differential equations (such as the **wave equation**).

However, sometimes when an imaginary or complex number arises in a calculation in physics or engineering, the only significance is that there is no real solution. For example, if a rock is thrown straight upwards and you proceed to calculate how much time it will take the rock to reach a particular height, if the answer is imaginary (and if your calculation is correct[3]), it is simply telling you that the rock wasn't thrown fast enough to ever reach that height. As another example, if a box is sliding down an inclined plane with friction and you proceed to calculate how fast the box will be moving when it reaches the bottom of the incline, if the answer is imaginary (and if your calculation is correct), it is telling you that there is so much friction that the box won't actually reach the bottom of the incline. Thus, an imaginary or complex number is sometimes just the way that mathematics tells you when something can't actually happen in the real world.

---

[3] However, most physics textbooks **don't** include problems where answers to motion problems are imaginary. So if you're solving a motion problem from a physics textbook and obtain an imaginary answer, the most likely explanation is that you made a mistake in your calculation.

Chapter 17 – Applications

Example 1. Use complex numbers to derive the trig identities below.
$$\cos(3\theta) = 4\cos^3\theta - 3\cos\theta \quad , \quad \sin(3\theta) = 3\sin\theta - 4\sin^3\theta$$
Use Euler's formula (Chapter 11) with an argument of 3θ.
$$e^{3i\theta} = \cos(3\theta) + i\sin(3\theta)$$
Use the rule $\left(e^{i\theta}\right)^a = e^{ai\theta}$ with $a = 3$. (As we learned in Chapter 16, we need to be careful when applying rules involving exponents. In this case the periodicity of the exponential function translates nicely into the periodicity of the trig functions, but since Chapter 16 demonstrated that the periodicity of complex numbers can lead to trouble, it's worth thinking each case through carefully.)
$$e^{3i\theta} = \left(e^{i\theta}\right)^3 = (\cos\theta + i\sin\theta)^3 = (\cos\theta + i\sin\theta)(\cos\theta + i\sin\theta)^2$$
$$= (\cos\theta + i\sin\theta)(\cos^2\theta + 2i\sin\theta\cos\theta - \sin^2\theta)$$
$$= \cos^3\theta + 2i\sin\theta\cos^2\theta - \sin^2\theta\cos\theta + i\sin\theta\cos^2\theta - 2\sin^2\theta\cos\theta - i\sin^3\theta$$
$$= \cos^3\theta + 3i\sin\theta\cos^2\theta - 3\sin^2\theta\cos\theta - i\sin^3\theta$$
Equate the above expression with $e^{3i\theta} = \cos(3\theta) + i\sin(3\theta)$ from the beginning of the solution. The real parts must be equal and the imaginary parts must also be equal, which gives two separate equations:
$$\cos(3\theta) = \cos^3\theta - 3\sin^2\theta\cos\theta \quad , \quad \sin(3\theta) = 3\sin\theta\cos^2\theta - \sin^3\theta$$
Since $\sin^2\theta + \cos^2\theta = 1$, it follows that $\sin^2\theta = 1 - \cos^2\theta$ and $\cos^2\theta = 1 - \sin^2\theta$.
$$\cos(3\theta) = \cos^3\theta - 3(1 - \cos^2\theta)\cos\theta \quad , \quad \sin(3\theta) = 3\sin\theta(1 - \sin^2\theta) - \sin^3\theta$$
$$\cos(3\theta) = \cos^3\theta - 3\cos\theta + 3\cos^3\theta \quad , \quad \sin(3\theta) = 3\sin\theta - 3\sin^3\theta - \sin^3\theta$$
$$\cos(3\theta) = 4\cos^3\theta - 3\cos\theta \quad , \quad \sin(3\theta) = 3\sin\theta - 4\sin^3\theta$$
Notice how we were able to derive two separate trig identities at the same time by using complex numbers.

Example 2. Find the eigenvalues for the matrix below.
$$\begin{pmatrix} 2 & 5 \\ -5 & 8 \end{pmatrix}$$
Identify $q = 2, r = 5, s = -5$, and $t = 8$. Use the formula $(q - \lambda)(t - \lambda) - rs = 0$.
$$(2 - \lambda)(8 - \lambda) - 5(-5) = 0$$
$$16 - 2\lambda - 8\lambda + \lambda^2 + 25 = 0$$
$$\lambda^2 - 10\lambda + 41 = 0$$
This is a quadratic equation with $a = 1, b = -10$, and $c = 41$. The discriminant (recall Chapter 9) for this quadratic is $b^2 - 4ac = (-10)^2 - 4(1)(41) = 100 - 164 = -64$. Alternatively, use the formula $(q - t)^2 + 4rs$ to find that the discriminant is equal to $(2 - 8)^2 + 4(5)(-5) = (-6)^2 - 100 = 36 - 100 = -64$. Either way, both eigenvalues are complex numbers because the discriminant is negative. Use the quadratic formula.

$$\lambda = \frac{-b \pm \sqrt{b^2 - 4ac}}{2a} = \frac{-(-10) \pm \sqrt{(-10)^2 - 4(1)(41)}}{2(1)} = \frac{10 \pm \sqrt{100 - 164}}{2}$$

$$\lambda = \frac{10 \pm \sqrt{-64}}{2} = \frac{10 \pm \sqrt{(64)(-1)}}{2} = \frac{10 \pm 8i}{2} = 5 \pm 4i$$

The two eigenvalues are $5 + 4i$ and $5 - 4i$.

Chapter 17 Problems

Directions: Use complex numbers to derive each trig identity.

(1) $\cos^4 \theta = \dfrac{3 + 4\cos(2\theta) + \cos(4\theta)}{8}$

(2) $\sin^4 \theta = \dfrac{3 - 4\cos(2\theta) + \cos(4\theta)}{8}$

(3) $\sin \alpha + \sin \beta = 2 \sin\left(\frac{\alpha+\beta}{2}\right) \cos\left(\frac{\alpha-\beta}{2}\right)$

(4) $\cos \alpha + \cos \beta = 2 \cos\left(\frac{\alpha+\beta}{2}\right) \cos\left(\frac{\alpha-\beta}{2}\right)$

Chapter 17 – Applications

Directions: Find the eigenvalues for each matrix.

(5)
$$\begin{pmatrix} 1 & \sqrt{2} \\ -\sqrt{2} & 3 \end{pmatrix}$$

(6)
$$\begin{pmatrix} 0 & -i \\ i & 0 \end{pmatrix}$$

Appendix: Cubic and Quartic Formulas

To use the cubic formula, first put the equation in the following **standard form**. If your cubic has a coefficient in front of z^3, you'll need to **divide both sides of the equation by that coefficient** before using the formulas below. For example, for the cubic equation $3x^3 + 6x^2 + 12x - 9 = 0$, first divide by 3 on both sides to get $x^3 + 2x^2 + 4x - 3 = 0$.

$$z^3 + az^2 + bz + c = 0$$

Next, calculate the following quantities:[1]

$$Q = \frac{3b - a^2}{9}, \quad R = \frac{9ab - 27c - 2a^3}{54}$$

$$S = \sqrt[3]{R + \sqrt{Q^3 + R^2}}, \quad T = \sqrt[3]{R - \sqrt{Q^3 + R^2}}$$

The solutions to the cubic equation are:

$$z_1 = S + T - \frac{a}{3}$$

$$z_2 = -\frac{S+T}{2} - \frac{a}{3} + \frac{\sqrt{3}}{2}i(S - T)$$

$$z_3 = -\frac{S+T}{2} - \frac{a}{3} - \frac{\sqrt{3}}{2}i(S - T)$$

If $Q^3 + R^2 < 0$ (meaning that $Q^3 < -R^2$), then $\sqrt{Q^3 + R^2}$ is imaginary (yet in this case the roots are real!) such that you will need to find cube roots of complex numbers (like we did in Chapter 13) in order to calculate S and T, as shown in Example 1.

The **discriminant** for the cubic is $Q^3 + R^2$. (This is the part under the square root in the formulas above.) If the **coefficients** of the polynomial are **real numbers**, then:

- If $Q^3 + R^2 < 0$, there are three real, distinct roots.
- If $Q^3 + R^2 = 0$, the roots are real and at least two are equal.
- If $Q^3 + R^2 > 0$, there are two complex roots and one real root, and the complex roots are complex conjugates of one another.

---

[1] Beware that not all texts use the same notation for the cubic and quartic formulas. For example, some texts use a_1, a_2, and a_3 instead of a, b, and c for the coefficients of the cubic.

Appendix – Cubic and Quartic Formulas

For the quartic, first put the equation in the following **standard form**. As mentioned for the cubic, if your cubic has a coefficient in front of z^4, you'll need to **divide both sides of the equation by that coefficient** before using the formulas below.
$$z^4 + ez^3 + fz^2 + gz + h = 0$$
Next, make the following cubic equation from the above coefficients:
$$y^3 - fy^2 + (eg - 4h)y + 4fh - g^2 - e^2h = 0$$
Observe that this cubic has $a = -f$, $b = eg - 4h$, and $c = 4fh - g^2 - e^2h$. Find a **real** root of this cubic and call it y_r. All four roots of the quartic equation can then be found by solving the **quadratic equation** below. Guess what! Then you need to use the quadratic formula from Chapter 9, as illustrated in Example 3. There are 4 possible combinations of the $\pm$ signs, which leads to 4 pairs of possible answers (since the quadratic formula gives a pair of solutions for each combination). Two combinations will produce two pairs of solutions to the quartic. To check if a solution of the quadratic below solves the quartic, plug it into the quartic equation.

$$z^2 + \frac{e \pm \sqrt{e^2 - 4f + 4y_r}}{2} z + \frac{y_r \pm \sqrt{y_r^2 - 4h}}{2} = 0$$

Example 1. Find all of the solutions to $5z^3 + 5z + 30 = 20z^2$.

Solution: To put this cubic in standard form, we need to subtract $20z^2$ from both sides and we also need to divide by 5 on both sides so that the coefficient of z^3 equals one. This gives us $z^3 - 4z^2 + z + 6 = 0$. Now identify $a = -4$, $b = 1$, and $c = 6$. Plug these values into the formulas for Q and R.

$$Q = \frac{3b - a^2}{9} = \frac{3(1) - (-4)^2}{9} = \frac{3 - 16}{9} = -\frac{13}{9}$$

$$R = \frac{9ab - 27c - 2a^3}{54} = \frac{9(-4)(1) - 27(6) - 2(-4)^3}{54} = \frac{-36 - 162 - 2(-64)}{54}$$

$$R = \frac{-198 + 128}{54} = -\frac{70}{54} = -\frac{35}{27}$$

The **discriminant** is $Q^3 + R^2 = \left(-\frac{13}{9}\right)^3 + \left(-\frac{35}{27}\right)^2 = -\frac{2197}{729} + \frac{1225}{729} = -\frac{972}{729} = -\frac{4}{3}$. Note that a negative number cubed is negative, as in $(-13)^3 = -2197$, whereas a negative number squared is positive, as in $(-35)^2 = 1225$. Since the discriminant is negative, there are three real, distinct roots. As discussed earlier in the chapter, when the discriminant is negative, $\sqrt{Q^3 + R^2}$ is imaginary, so we will need to find cube roots of a complex number (like we did in Chapter 13) in order to calculate S and T.

The square root of the discriminant is $\sqrt{Q^3 + R^2} = \sqrt{-\frac{4}{3}} = \sqrt{(-1)\left(\frac{4}{3}\right)} \approx 1.1547i$. Plug $R = -\frac{35}{27} \approx -1.2963$ and $\sqrt{Q^3 + R^2} \approx 1.1547i$ into the formulas for S and T.

$$S = \sqrt[3]{R + \sqrt{Q^3 + R^2}} \approx \sqrt[3]{-1.2963 + 1.1547i}$$

$$T \approx \sqrt[3]{R - \sqrt{Q^3 + R^2}} \approx \sqrt[3]{-1.2963 - 1.1547i}$$

Now we will find the cube roots of these complex numbers following the strategy from Chapter 13. First, we'll do this for S.

$$\text{modulus} \approx \sqrt{(-1.2963)^2 + 1.1547^2} \approx \sqrt{1.6804 + 1.3333} \approx 1.736$$

$$\text{argument} \approx \tan^{-1}\left(\frac{1.1547}{-1.2963}\right) \approx \tan^{-1}(-1.5429) \approx -0.72769 + \pi \approx 2.4139 \text{ rad}$$

Note that we worked this out using **radians**. As discussed in Chapter 7, add π radians to the calculator's answer when the real part is negative in order to put the angle in the correct quadrant, which in this case is Quadrant II.

$$S \approx \sqrt[3]{-1.2963 + 1.1547i} \approx \sqrt[3]{1.736e^{2.4139i}} \approx \left(1.736e^{2.4139i}\right)^{1/3} \approx 1.736^{1/3}e^{2.4139i/3}$$

$$S \approx 1.2018e^{0.8046i} = 1.2018\cos 0.8046 + 1.2018i \sin 0.8046 \approx 0.8333 + 0.8660i$$

We don't really need to repeat all of those steps for T (but you can if you want). The only difference for T is that the imaginary part is -1.1547 instead of $+1.1547$. This will leave modulus unchanged and change the argument to -2.4139 rad (which is equivalent to 3.8693 rad, since they differ by 2π). All this does is change the sign of 0.8046 rad to -0.8046 rad (which is equivalent to 5.4786 rad) in the last step:

$$T \approx 1.2018\cos(-0.8046) + 1.2018i\sin(-0.8046) \approx 0.8333 - 0.8660i$$

It's convenient to calculate the following before using the final equations:

$$S + T \approx 0.8333 + 0.8660i + 0.8333 - 0.8660i \approx 1.67$$

$$-\frac{S+T}{2} \approx -\frac{0.8333 + 0.8660i + 0.8333 - 0.8660i}{2} \approx -0.83$$

$$\frac{\sqrt{3}}{2}i(S - T) \approx \sqrt{3}\frac{0.8333 + 0.8660i - 0.8333 + 0.8660i}{2}i$$

Appendix – Cubic and Quartic Formulas

$$\frac{\sqrt{3}}{2}i(S-T) \approx \sqrt{3}(0.8660)i^2 = \sqrt{3}(0.8660)(-1) \approx -1.50$$

Finally, plug these values into the formulas for the solutions to the cubic equation.

$$z_1 = S + T - \frac{a}{3} \approx 1.67 - \frac{(-4)}{3} \approx 1.67 + 1.33 = 3$$

$$z_2 = -\frac{S+T}{2} - \frac{a}{3} + \frac{\sqrt{3}}{2}i(S-T) = -0.83 - \frac{(-4)}{3} - 1.50 \approx -2.33 + 1.33 = -1$$

$$z_3 = -\frac{S+T}{2} - \frac{a}{3} - \frac{\sqrt{3}}{2}i(S-T) = -0.83 - \frac{(-4)}{3} + 1.50 \approx 0.67 + 1.33 = 2$$

Did you notice? The given polynomial has purely **real** coefficients and all three of the solutions are purely **real**, yet in order to solve the problem we had to take the cube root of a **complex** number. Our knowledge of complex numbers was applied to solve a purely real equation with purely real solutions. Let that soak in.

Check: Plug each answer into the original equation to check it.
$5z^3 + 5z + 30 = 5(3)^3 + 5(3) + 30 = 5(27) + 15 + 30 = 135 + 45 = 180$ agrees with $20z^2 = 20(3)^2 = 20(9) = 180$ ✓
$5z^3 + 5z + 30 = 5(-1)^3 + 5(-1) + 30 = 5(-1) - 5 + 30 = -5 - 5 + 30 = 20$ agrees with $20z^2 = 20(-1)^2 = 20(1) = 20$ ✓
$5z^3 + 5z + 30 = 5(2)^3 + 5(2) + 30 = 5(8) + 10 + 30 = 40 + 40 = 80$ agrees with $20z^2 = 20(2)^2 = 20(4) = 80$ ✓

Example 2. Find all of the solutions to $2z^3 + 110z = 22z^2 + 250$.
Solution: To put this cubic in standard form, we need to subtract $22z^2$ and 250 from both sides and we also need to divide by 2 on both sides so that the coefficient of z^3 equals one. This gives us $z^3 - 11z^2 + 55z - 125 = 0$. Now identify $a = -11, b = 55$, and $c = -125$. Plug these values into the formulas for Q and R.

$$Q = \frac{3b - a^2}{9} = \frac{3(55) - (-11)^2}{9} = \frac{165 - 121}{9} = \frac{44}{9}$$

$$R = \frac{9ab - 27c - 2a^3}{54} = \frac{9(-11)(55) - 27(-125) - 2(-11)^3}{54}$$

$$R = \frac{-5445 + 3375 - 2(-1331)}{54} = \frac{-2070 + 2662}{54} = \frac{592}{54} = \frac{296}{27}$$

The **discriminant** is $Q^3 + R^2 = \left(\frac{44}{9}\right)^3 + \left(\frac{296}{27}\right)^2 = \frac{85{,}184}{729} + \frac{87{,}616}{729} = \frac{172{,}800}{729} = \frac{6400}{27} \approx 237$. Since the discriminant is positive, there are two complex roots (which are complex conjugates of one another) and one real root. Plug $R = \frac{296}{27} \approx 10.96$ and $\sqrt{Q^3 + R^2} \approx \sqrt{237} \approx 15.39$ into the formulas for S and T.

$$S = \sqrt[3]{R + \sqrt{Q^3 + R^2}} \approx \sqrt[3]{10.96 + 15.39} \approx \sqrt[3]{26.35} \approx 2.98$$

$$T \approx \sqrt[3]{R - \sqrt{Q^3 + R^2}} \approx \sqrt[3]{10.96 - 15.39} \approx \sqrt[3]{-4.43} \approx -1.64$$

Finally, plug these values into the formulas for the solutions to the cubic equation.

$$z_1 = S + T - \frac{a}{3} \approx 2.98 - 1.64 - \frac{(-11)}{3} \approx 1.34 + 3.67 \approx 5$$

$$z_2 = -\frac{S+T}{2} - \frac{a}{3} + \frac{\sqrt{3}}{2}i(S - T) \approx -\frac{2.98 - 1.64}{2} - \frac{(-11)}{3} + \frac{\sqrt{3}}{2}i[2.98 - (-1.64)]$$

$$z_2 \approx -0.67 + 3.67 + \frac{4.62\sqrt{3}}{2}i \approx 3 + 4i$$

$$z_3 \approx 3 - 4i$$

Note: The two complex solutions (z_2 and z_3) are complex conjugates of one another. **Compare and contrast**: Recall that Example 1 had real coefficients and also real solutions, yet required finding the cube root of a complex number. Observe that we didn't need to find the root of a complex number in Example 2, which had complex numbers as solutions. It was actually easier to solve the problem where the answers included complex numbers than when all of the answers were real. Think about that. Check: Plug each answer into the original equation to check it.
$2z^3 + 110z = 2(5)^3 + 110(5) = 2(125) + 550 = 250 + 550 = 800$ agrees with
$22z^2 + 250 = 22(5)^2 + 250 = 22(25) + 250 = 550 + 250 = 800$ ✓
$2z^3 + 110z = 2(3 + 4i)^3 + 110(3 + 4i) = 2(3 + 4i)(9 + 24i - 16) + 330 + 440i$
$= (6 + 8i)(-7 + 24i) + 330 + 440i = -42 + 144i - 56i - 192 + 330 + 440i$
$= 96 + 528i$ agrees with $22z^2 + 250 = 22(3 + 4i)^2 + 250 = 22(9 + 24i - 16) + 250$
$= 198 + 528i - 352 + 250 = 96 - 528i$ ✓

Appendix – Cubic and Quartic Formulas

Example 3. Find all of the solutions to $36z^4 + 6z = 24z^3 + 5z^2 + 1$.

Solution: To put this quartic in standard form, we need to move all of the terms to the same side and we also need to divide by 36 on both sides so that the coefficient of z^4 equals one. This gives us $z^4 - \frac{2}{3}z^3 - \frac{5}{36}z^2 + \frac{1}{6}z - \frac{1}{36} = 0$. Now identify $e = -\frac{2}{3}$, $f = -\frac{5}{36}$, $g = \frac{1}{6}$, and $h = -\frac{1}{36}$. To solve a quartic, first we need to solve a corresponding cubic equation. Use the formulas that relate e, f, g, and h to a, b, and c.

$$a = -f = -\left(-\frac{5}{36}\right) = \frac{5}{36} \quad , \quad b = eg - 4h = \left(-\frac{2}{3}\right)\left(\frac{1}{6}\right) - 4\left(-\frac{1}{36}\right) = -\frac{1}{9} + \frac{1}{9} = 0$$

$$c = 4fh - g^2 - e^2h = 4\left(-\frac{5}{36}\right)\left(-\frac{1}{36}\right) - \left(\frac{1}{6}\right)^2 - \left(-\frac{2}{3}\right)^2\left(-\frac{1}{36}\right)$$

$$c = \frac{20}{1296} - \frac{1}{36} - \frac{4}{9}\left(-\frac{1}{36}\right) = \frac{20}{1296} - \frac{1}{36} + \frac{1}{81} = \frac{20}{1296} - \frac{36}{1296} + \frac{16}{1296} = 0$$

This makes the following cubic equation:

$$y^3 + ay^2 + by + c = 0 \rightarrow y^3 + \frac{5}{36}y^2 = 0$$

Two terms happen to be zero because b and c happen to be zero. Ordinarily, in this step you would plug the values for a, b, and c into the formulas for Q and R to solve the cubic, but since b and c are both zero in this example, we don't need the cubic formula to solve for y. Just factor out y^2 in the above equation.

$$y^2\left(y + \frac{5}{36}\right) = 0$$

Either $y^2 = 0$, which is a double root with $y = 0$, or $y + \frac{5}{36} = 0$, which gives us the third root $y = -\frac{5}{36}$. We only need one real root. We'll choose $y = 0$ because that makes the calculation simpler. Call the real root $y_r = 0$ (where the subscript stands for "real"). Now we need to solve the following quadratic equation.

$$z^2 + \frac{e \pm \sqrt{e^2 - 4f + 4y_r}}{2}z + \frac{y_r \pm \sqrt{y_r^2 - 4h}}{2} = 0$$

Plug $e = -\frac{2}{3}$, $f = -\frac{5}{36}$, $g = \frac{1}{6}$, $h = -\frac{1}{36}$, and $y_r = 0$ into the previous equation.

$$z^2 + \frac{-\frac{2}{3} \pm \sqrt{\left(-\frac{2}{3}\right)^2 - 4\left(-\frac{5}{36}\right) + 4(0)}}{2}z + \frac{0 \pm \sqrt{0^2 - 4\left(-\frac{1}{36}\right)}}{2} = 0$$

$$z^2 + \frac{-\frac{2}{3} \pm \sqrt{\frac{4}{9} + \frac{5}{9} + 0}}{2} z \pm \frac{\sqrt{0 + \frac{1}{9}}}{2} = 0$$

$$z^2 + \frac{-\frac{2}{3} \pm \sqrt{1}}{2} z \pm \frac{\sqrt{\frac{1}{9}}}{2} = 0$$

$$z^2 + \frac{-\frac{2}{3} \pm 1}{2} z \pm \frac{1}{6} = 0$$

The ± signs give 4 possible combinations:

$$z^2 + \frac{-\frac{2}{3}+1}{2} z + \frac{1}{6} = 0 \text{ or } z^2 + \frac{-\frac{2}{3}+1}{2} z - \frac{1}{6} = 0 \text{ or } z^2 + \frac{-\frac{2}{3}-1}{2} z + \frac{1}{6} = 0 \text{ or } z^2 + \frac{-\frac{2}{3}-1}{2} z - \frac{1}{6} = 0$$

$$z^2 + \frac{1/3}{2} z + \frac{1}{6} = 0 \text{ or } z^2 + \frac{1/3}{2} z - \frac{1}{6} = 0 \text{ or } z^2 + \frac{\left(-\frac{5}{3}\right)}{2} z + \frac{1}{6} = 0 \text{ or } z^2 + \frac{\left(-\frac{5}{3}\right)}{2} z - \frac{1}{6} = 0$$

$$z^2 + \frac{1}{6} z + \frac{1}{6} = 0 \text{ or } z^2 + \frac{1}{6} z - \frac{1}{6} = 0 \text{ or } z^2 - \frac{5}{6} z + \frac{1}{6} = 0 \text{ or } z^2 - \frac{5}{6} z - \frac{1}{6} = 0$$

In this example, it turns out that the combinations $z^2 + \frac{1}{6} z - \frac{1}{6} = 0$ and $z^2 - \frac{5}{6} z + \frac{1}{6} = 0$ lead to correct solutions for the quartic. The "check" following the solution shows that these result in the correct answers. We will focus on these two combinations, since the others don't lead to correct answers for the quartic.

$$z = \frac{-\frac{1}{6} \pm \sqrt{\left(\frac{1}{6}\right)^2 - 4(1)\left(-\frac{1}{6}\right)}}{2(1)} \text{ or } z = \frac{-\left(-\frac{5}{6}\right) \pm \sqrt{\left(-\frac{5}{6}\right)^2 - 4(1)\left(\frac{1}{6}\right)}}{2(1)}$$

$$z = \frac{-\frac{1}{6} \pm \sqrt{\frac{1}{36} + \frac{2}{3}}}{2} \text{ or } z = \frac{\frac{5}{6} \pm \sqrt{\frac{25}{36} - \frac{2}{3}}}{2}$$

$$z = \frac{-\frac{1}{6} \pm \sqrt{\frac{1}{36} + \frac{24}{36}}}{2} \text{ or } z = \frac{\frac{5}{6} \pm \sqrt{\frac{25}{36} - \frac{24}{36}}}{2}$$

$$z = \frac{-\frac{1}{6} \pm \sqrt{\frac{25}{36}}}{2} \text{ or } z = \frac{\frac{5}{6} \pm \sqrt{\frac{1}{36}}}{2}$$

$$z = \frac{-\frac{1}{6} \pm \frac{5}{6}}{2} \text{ or } z = \frac{\frac{5}{6} \pm \frac{1}{6}}{2}$$

$$z = -\frac{1}{12} \pm \frac{5}{12} \text{ or } z = \frac{5}{12} \pm \frac{1}{12}$$

$$z = -\frac{1}{12} - \frac{5}{12} \text{ or } z = -\frac{1}{12} + \frac{5}{12} \text{ or } z = \frac{5}{12} - \frac{1}{12} \text{ or } z = \frac{5}{12} + \frac{1}{12}$$

$$z = -\frac{1}{2} \text{ or } z = \frac{1}{3} \text{ or } z = \frac{1}{3} \text{ or } z = \frac{1}{2}$$

Note: Since $z = \frac{1}{3}$ is a double root, there are only 3 distinct answers to this quartic.

Check: Plug each answer into the original equation to check it.

$36z^4 + 6z = 36\left(-\frac{1}{2}\right)^4 + 6\left(-\frac{1}{2}\right) = \frac{36}{16} - \frac{6}{2} = \frac{9}{4} - \frac{12}{4} = -\frac{3}{4}$ agrees with

$24z^3 + 5z^2 + 1 = 24\left(-\frac{1}{2}\right)^3 + 5\left(-\frac{1}{2}\right)^2 + 1 = -\frac{24}{8} + \frac{5}{4} + 1 = -\frac{12}{4} + \frac{5}{4} + \frac{4}{4} = -\frac{3}{4}$ ✓

$36z^4 + 6z = 36\left(\frac{1}{3}\right)^4 + 6\left(\frac{1}{3}\right) = \frac{36}{81} + \frac{6}{3} = \frac{4}{9} + \frac{18}{9} = \frac{22}{9}$ agrees with

$24z^3 + 5z^2 + 1 = 24\left(\frac{1}{3}\right)^3 + 5\left(\frac{1}{3}\right)^2 + 1 = \frac{24}{27} + \frac{5}{9} + 1 = \frac{8}{9} + \frac{5}{9} + \frac{9}{9} = \frac{22}{9}$ ✓

$36z^4 + 6z = 36\left(\frac{1}{2}\right)^4 + 6\left(\frac{1}{2}\right) = \frac{36}{16} + \frac{6}{2} = \frac{9}{4} + \frac{12}{4} = \frac{21}{4}$ agrees with

$24z^3 + 5z^2 + 1 = 24\left(\frac{1}{2}\right)^3 + 5\left(\frac{1}{2}\right)^2 + 1 = \frac{24}{8} + \frac{5}{4} + 1 = \frac{12}{4} + \frac{5}{4} + \frac{4}{4} = \frac{21}{4}$ ✓

Appendix Problems

Directions: Use the cubic or quartic formulas to find all of the solutions to each equation.

(1) $2z^3 + 252z = 42z^2 + 432$

(2) $5z^3 + 5z + 170 = 30z^2$

(3) $\frac{z^3}{3} + 9z = 5z^2 - 81$

(4) $3z^4 + 4z^3 + 112z = 64z^2 + 48$

(5) $288 - 76z^2 + z^4 = 24z - 6z^3$

(6) $\frac{z^4}{4} - z^3 + z^2 = 1$

Complex Numbers Essentials Math Workbook with Answers

Chapter 1

(1) $i^{19} = i^{16}i^3 = (1)(-i) = -i$

Notes: 19 ÷ 4 has a remainder of 3 (4 × 4 = 16 and 19 − 16 = 3).

$$i^3 = i^7 = i^{11} = i^{15} = i^{19} = \cdots = -i$$

(2) $i^{56} = (i^4)^{14} = (1)^{14} = 1$

Notes: 56 ÷ 4 = 14 has no remainder; 56 is a multiple of 4.

$$i^4 = i^8 = i^{12} = i^{16} = i^{20} = \cdots = 1$$

(3) $i^{33} = i^{32}i^1 = (1)(i) = i$

Notes: 33 ÷ 4 has a remainder of 1 (8 × 4 = 32 and 33 − 32 = 1).

$$i^1 = i^5 = i^9 = i^{13} = i^{17} = \cdots = i$$

(4) $i^{2002} = i^{2000}i^2 = (1)(-1) = -1$

Notes: 2002 ÷ 4 has a remainder of 2 (500 × 4 = 2000 and 2002 − 2000 = 2).

$$i^2 = i^6 = i^{10} = i^{14} = i^{18} = \cdots = -1$$

(5) $(-3i)^3 = (-1)^3(3)^3 i^3 = (-1)(27)(-i) = 27i$

Notes: The minus signs from $(-1)^3$ and from $i^3 = -i$ cancel: $(-1)(-1) = 1$. We used the rule from algebra that $(x^m)^n = x^{mn}$.

(6) $(4i)^4 = 4^4 i^4 = (256)(1) = 256$

(7) $\frac{1}{i^3} = \frac{1}{-i} = -\frac{1}{i} = -\frac{1}{i}\frac{i}{i} = -\frac{i}{i^2} = -\frac{i}{-1} = i$

Notes: We multiplied by $\frac{i}{i}$. The minus sign from $i^2 = -1$ cancels with the overall minus sign.

Alternate solution: $\frac{1}{i^3} = \frac{1}{i^3}\frac{i}{i} = \frac{i}{i^4} = \frac{i}{1} = i$.

(8) $\left(\frac{2}{i}\right)^6 = \frac{2^6}{i^6} = \frac{64}{i^4 i^2} = \frac{64}{(1)(-1)} = \frac{64}{-1} = -64$

Alternate solution: $\left(\frac{2}{i}\right)^6 = \left(\frac{2}{i}\frac{i}{i}\right)^6 = \left(\frac{2i}{i^2}\right)^6 = \left(\frac{2i}{-1}\right)^6 = (-2i)^6 = (64)i^2 = -64$

(9) $\left(-\frac{5}{i}\right)^4 = \frac{(-5)^4}{i^4} = \frac{625}{1} = 625$

Alternate solution: $\left(-\frac{5}{i}\right)^4 = \left(-\frac{5}{i}\frac{i}{i}\right)^4 = \left(-\frac{5i}{i^2}\right)^4 = \left(-\frac{5i}{-1}\right)^4 = (5i)^4 = 5^4 i^4 = 625$

Answer Key

Chapter 2

(1) $2i^2 + 3i^3 + 5i^5 + 8i^8 + 13i^{13} = 2(-1) + 3(-i) + 5(i) + 8(1) + 13(i)$
$= -2 - 3i + 5i + 8 + 13i = (-2 + 8) + (-3i + 5i + 13i) = 6 + 15i$ (complex)
Notes: $i^2 = -1, i^3 = -i, i^5 = i^4 i^1 = i, i^8 = (i^4)^2 = 1, i^{13} = i^{12} i^1 = (i^4)^3 i = i$.
Recall that $i^4 = 1$.

(2) $7i^{15} + 5i^{27} - 3i^{49} - i^{75} = 7i^{12} i^3 + 5i^{24} i^3 - 3i^{48} i^1 - i^{72} i^3$
$= 7(i^4)^3(-i) + 5(i^4)^6(-i) - 3(i^4)^{12} i - (i^4)^{18}(-i) = -7i - 5i - 3i - (-i)$
$= -15i + i = -14i$ (purely imaginary)
Note: The 'simple' solution is to realize that any odd power of i is imaginary ($i^3 = -i, i^5 = i, i^7 = -i, i^9 = i, i^{11} = -i$, etc.), whereas any even power is real.

(3) $(2i)^8 - (3i)^6 + (4i)^4 - (6i)^2 = 2^8 i^8 - 3^6 i^6 + 4^4 i^4 - 6^2 i^2$
$= 256(1) - 729(-1) + 256(1) - 36(-1) = 256 + 729 + 256 + 36 = 1277$ (real)
Notes: The 'simple' solution is that any even power of i is real ($i^2 = -1, i^4 = 1, i^6 = -1, i^8 = 1$, etc.). Recall the rule from algebra that $(x^m)^n = x^{mn}$.

(4) $\frac{i}{8} + \left(\frac{i}{4}\right)^2 + \left(\frac{i}{2}\right)^3 = \frac{i}{8} + \frac{i^2}{4^2} + \frac{i^3}{2^3} = \frac{i}{8} + \frac{-1}{16} + \frac{-i}{8} = \frac{i}{8} - \frac{1}{16} - \frac{i}{8} = -\frac{1}{16}$ (real)

(5) $\frac{1}{2} - \frac{2}{i} = \frac{1}{2} - \frac{2i}{ii} = \frac{1}{2} - \frac{2i}{i^2} = \frac{1}{2} - \left(\frac{2i}{-1}\right) = \frac{1}{2} - (-2i) = \frac{1}{2} + 2i$ (complex)

(6) $\frac{5}{i^5} + \frac{3}{i^3} - \frac{1}{i} = \frac{5}{i} + \frac{3}{-i} - \frac{1}{i} = \frac{5}{i} - \frac{3}{i} - \frac{1}{i} = \frac{1}{i}(5 - 3 - 1) = \frac{1}{i} = \frac{1 i}{i i} = \frac{i}{i^2} = \frac{i}{-1} = -i$
(purely imaginary)
Note: Compare with Exercise 2. These powers include i^{-5}, i^{-3}, and i^{-1}.

(7) $i^3 \sqrt{3} + 2i^4 + 3i^5 + 4i^6 \sqrt{6} = (-i)\sqrt{3} + 2(1) + 3(i) + 4(-1)\sqrt{6}$
$= -i\sqrt{3} + 2 + 3i - 4\sqrt{6} = (2 - 4\sqrt{6}) + (3 - \sqrt{3})i$ (complex)
Note: The term $2 - 4\sqrt{6}$ is real, whereas the term $(3 - \sqrt{3})i$ is imaginary.

(8) $9i^9 + i^7 \sqrt{7} - 5i^5 = 9(i) + i^3 \sqrt{7} - 5(i) = 9i + (-i)\sqrt{7} - 5i = 4i - i\sqrt{7}$
$= (4 - \sqrt{7})i$ (purely imaginary)
Notes: This is very much like Exercise 2; every term has an odd power of i. The terms $4i$ and $-i\sqrt{7}$ are both purely imaginary; there are no real terms.

Chapter 3

(1) $\overline{4+2i} = 4-2i$

(2) $\overline{7-5i} = 7+5i$

(3) $\overline{i+1} = -i+1$ Alternate answer: $1-i$

(4) $\overline{-11+6i} = -11-6i$ Alternate answer: $-(11+6i)$

Note: Don't change the sign of the real part.

(5) $\overline{8i+15} = \overline{15+8i} = 15-8i$ Alternate answer: $-8i+15$

(6) $\overline{(5i)^3} = \overline{5^3 i^3} = \overline{(125)(-i)} = \overline{-125i} = 125i$ Notes: There isn't any real part. Simplify **before** reversing the sign of the imaginary part.

(7) $\overline{-3-9i} = -3+9i$ Note: Don't change the sign of the real part.

(8) $\overline{3+\frac{6}{i}} = \overline{3+\frac{6\,i}{i\,i}} = \overline{3+\frac{6i}{i^2}} = \overline{3+\frac{6i}{-1}} = \overline{3-6i} = 3+6i$

(9) $\overline{i^{22}+2i^{17}-3i^8} = \overline{i^{20}i^2+2i^{16}i^1-3(i^4)^2} = \overline{(i^4)^5(-1)+2(i^4)^4 i-3(1)^2}$
$= \overline{(1)^5(-1)+2(1)^4 i-3} = \overline{-1+2i-3} = \overline{-4+2i} = -4-2i$

(10) $(6-3i)\overline{6-3i} = 6^2+(-3)^2 = 36+9 = 45$

(11) $(i+2)\overline{i+2} = 1^2+2^2 = 1+4 = 5$

Alternate solution: $(i+2)\overline{i+2} = (i+2)(-i+2) = -i^2+2i-2i+4 = -i^2+4$
$= -(-1)+4 = 1+4 = 5$

(12) $i\bar{i} = i(-i) = -i^2 = -(-1) = 1$

Tip: $(x+iy)\overline{x+iy} = (x+iy)(x-iy)$ is always **nonnegative** (if x and y are real).

(13) $(5-12i)\overline{5-12i} = 5^2+(-12)^2 = 25+144 = 169$

(14) $(-7+8i)\overline{-7+8i} = (-7)^2+8^2 = 49+64 = 113$

(15) $(-9-4i)\overline{-9-4i} = (-9)^2+(-4)^2 = 81+16 = 97$

(16) $i^{12}-2i^9+3i^6 = (i^4)^3-2i^8 i^1+3i^4 i^2 = 1-2i-3 = -2-2i$

$(i^{12}-2i^9+3i^6)\overline{i^{12}-2i^9+3i^6} = (-2-2i)\overline{-2-2i} = (-2)^2+(-2)^2 = 4+4 = 8$

(17) $2+\frac{1}{i} = 2+\frac{1\,i}{i\,i} = 2+\frac{i}{i^2} = 2+\frac{i}{-1} = 2-i$

$\left(2+\frac{1}{i}\right)\overline{2+\frac{1}{i}} = (2-i)\overline{2-i} = 2^2+(-1)^2 = 4+1 = 5$ Tip: All terms are **nonnegative**.

Answer Key

Chapter 4

(1) $10 + 6i + 12 - 4i = (10 + 12) + (6i - 4i) = 22 + 2i$

(2) $(2 + 7i) - (5 + 3i) = 2 + 7i - 5 - 3i = (2 - 5) + (7i - 3i) = -3 + 4i$

(3) $(-4 + 9i) + (-6 + 8i) = -4 + 9i - 6 + 8i = (-4 - 6) + (9i + 8i) = -10 + 17i$

Note: There is a plus sign (not a minus sign) between the two sets of parentheses. It's an addition problem, not a subtraction problem.

(4) $11 + 3i - (6 - 9i) = 11 + 3i - 6 - (-9i) = 11 + 3i - 6 + 9i$
$= (11 - 6) + (3i + 9i) = 5 + 12i$

(5) $(15 - 10i) - (7 + 6i) = 15 - 10i - 7 - 6i = (15 - 7) + (-10i - 6i) = 8 - 16i$

(6) $24 - 18i + (16 - 12i) = 24 - 18i + 16 - 12i = (24 + 16) + (-18i - 12i)$
$= 40 - 30i$

Note: There is a plus sign (not a minus sign) between the two sets of parentheses.

(7) $21 - 12i - (21 + 12i) = 21 - 12i - 21 - 12i = (21 - 21) + (-12i - 12i) = -24i$

(8) $(17 - 11i) - (-8 + 4i) = 17 - 11i - (-8) - 4i = 17 - 11i + 8 - 4i$
$= (17 + 8) + (-11i - 4i) = 25 - 15i$

(9) $(32 + 48i) + (-17 + 24i) = 32 + 48i - 17 + 24i = (32 - 17) + (48i + 24i)$
$= 15 + 72i$

Note: There is a plus sign (not a minus sign) between the two sets of parentheses.

(10) $(-60 + 40i) - (-20 + 80i) = -60 + 40i - (-20) - 80i = -60 + 40i + 20 - 80i$
$= (-60 + 20) + (40i - 80i) = -40 - 40i$

(11) $(8 + 4i)(5 + 6i) = 8(5) + 8(6i) + 4i(5) + 4i(6i) = 40 + 48i + 20i + 24i^2$
$= 40 + 68i + 24(-1) = 40 + 68i - 24 = 16 + 68i$

(12) $(9 + 5i)(4 - 7i) = 9(4) + 9(-7i) + 5i(4) + 5i(-7i) = 36 - 63i + 20i - 35i^2$
$= 36 - 43i - 35(-1) = 36 - 43i + 35 = 71 - 43i$

(13) $(6 - 7i)(8 - 3i) = 6(8) + 6(-3i) - 7i(8) - 7i(-3i) = 48 - 18i - 56i + 21i^2$
$= 48 - 74i + 21(-1) = 48 - 74i - 21 = 27 - 74i$

(14) $(7 - 5i)(5 + 9i) = 7(5) + 7(9i) - 5i(5) - 5i(9i) = 35 + 63i - 25i - 45i^2$
$= 35 + 38i - 45(-1) = 35 + 38i + 45 = 80 + 38i$

(15) $(3 - 9i)\overline{8 - 4i} = (3 - 9i)(8 + 4i) = 3(8) + 3(4i) - 9i(8) - 9i(4i)$
$= 24 + 12i - 72i - 36i^2 = 24 - 60i - 36(-1) = 24 - 60i + 36 = 60 - 60i$

Note: Recall from Chapter 3 that $\overline{8 - 4i} = 8 + 4i$ is the complex conjugate of $8 - 4i$.

(16) $(8+5i)^2 = (8+5i)(8+5i) = 8(8) + 8(5i) + 5i(8) + 5i(5i)$
$= 64 + 40i + 40i + 25i^2 = 64 + 80i + 25(-1) = 64 + 80i - 25 = 39 + 80i$

(17) First find $(2-i)^2 = (2-i)(2-i) = 2(2) + 2(-i) - i(2) - i(-i)$
$(2-i)^2 = 4 - 2i - 2i + i^2 = 4 - 4i - 1 = 3 - 4i$ Now multiply this by $(2-i)$.
$(2-i)^3 = (2-i)^2(2-i) = (3-4i)(2-i) = 3(2) + 3(-i) - 4i(2) - 4i(-i)$
$(2-i)^3 = 6 - 3i - 8i + 4i^2 = 6 - 11i + 4(-1) = 6 - 11i - 4 = 2 - 11i$

(18) Multiply the answer to Exercise 17 by $(2-i)$.
$(2-i)^4 = (2-i)^3(2-i) = (2-11i)(2-i) = 2(2) + 2(-i) - 11i(2) - 11i(-i)$
$= 4 - 2i - 22i + 11i^2 = 4 - 24i + 11(-1) = 4 - 24i - 11 = -7 - 24i$

(19) $(4-8i) \div (2+5i) = \frac{4-8i}{2+5i} = \frac{4-8i}{2+5i}\left(\frac{2-5i}{2-5i}\right) = \frac{4(2)+4(-5i)-8i(2)-8i(-5i)}{2^2+5^2}$
$= \frac{8-20i-16i+40i^2}{4+25} = \frac{8-36i+40(-1)}{29} = \frac{8-36i-40}{29} = \frac{-32-36i}{29} = -\frac{32}{29} - \frac{36}{29}i$
Check the answer: $\frac{-32-36i}{29}(2+5i) = \frac{-32(2)-32(5i)-36i(2)-36i(5i)}{29} = \frac{-64-160i-72i-180i^2}{29}$
$= \frac{-64-232i-180(-1)}{29} = \frac{-64-232i+180}{29} = \frac{116-232i}{29} = \frac{116}{29} - \frac{232i}{29} = 4 - 8i$

(20) $(15-25i) \div (2-i) = \frac{15-25i}{2-i} = \frac{15-25i}{2-i}\left(\frac{2+i}{2+i}\right) = \frac{15(2)+15(i)-25i(2)-25i(i)}{2^2+1^2}$
$= \frac{30+15i-50i-25i^2}{4+1} = \frac{30-35i-25(-1)}{5} = \frac{30-35i+25}{5} = \frac{55-35i}{5} = \frac{5(11-7i)}{5} = 11 - 7i$
Check the answer: $(11-7i)(2-i) = 11(2) + 11(-i) - 7i(2) - 7i(-i)$
$= 22 - 11i - 14i + 7i^2 = 22 - 25i + 7(-1) = 22 - 25i - 7 = 15 - 25i$

(21) $\frac{6}{1+i} = \frac{6}{1+i}\left(\frac{1-i}{1-i}\right) = \frac{6-6i}{1^2+1^2} = \frac{6-6i}{1+1} = \frac{6-6i}{2} = \frac{6}{2} - \frac{6i}{2} = 3 - 3i$
Check the answer: $(3-3i)(1+i) = 3(1) + 3(i) - 3i(1) - 3i(i) = 3 + 3i - 3i - 3i^2$
$= 3 + 0 - 3(-1) = 3 + 3 = 6$

(22) $\frac{5+10i}{4-3i} = \frac{5+10i}{4-3i}\left(\frac{4+3i}{4+3i}\right) = \frac{5(4)+5(3i)+10i(4)+10i(3i)}{4^2+3^2} = \frac{20+15i+40i+30i^2}{16+9} = \frac{20+55i+30(-1)}{25}$
$= \frac{20+55i-30}{25} = \frac{-10+55i}{25} = \frac{5(-2+11i)}{5(5)} = \frac{-2+11i}{5} = -\frac{2}{5} + \frac{11}{5}i$
Check the answer: $\frac{-2+11i}{5}(4-3i) = \frac{-2(4)-2(-3i)+11i(4)+11i(-3i)}{5} = \frac{-8+6i+44i-33i^2}{5}$
$= \frac{-8+50i-33(-1)}{5} = \frac{-8+50i+33}{5} = \frac{25+50i}{5} = \frac{25}{5} + \frac{50i}{5} = 5 + 10i$

(23) $\frac{-120+240i}{-2+4i} = \frac{-120+240i}{-2+4i}\left(\frac{-2-4i}{-2-4i}\right) = \frac{-120(-2)-120(-4i)+240i(-2)+240i(-4i)}{(-2)^2+4^2}$

Answer Key

$$= \frac{240+480i-480i-960i^2}{4+16} = \frac{240+0-960(-1)}{20} = \frac{240+960}{20} = \frac{1200}{20} = 60$$

Alternate solution: $\frac{-120+240i}{-2+4i} = \frac{60(-2+4i)}{-2+4i} = 60$

Check the answer: $60(-2+4i) = -120+240i$

(24) $\frac{(8+i)(4-i)}{(3-2i)^2} = \frac{8(4)+8(-i)+i(4)+i(-i)}{(3-2i)(3-2i)} = \frac{32-8i+4i-i^2}{3(3)+3(-2i)-2i(3)-2i(-2i)} = \frac{32-4i-(-1)}{9-6i-6i+4i^2} = \frac{32-4i+1}{9-12i+4(-1)}$

$= \frac{33-4i}{9-12i-4} = \frac{33-4i}{5-12i} = \frac{33-4i}{5-12i}\left(\frac{5+12i}{5+12i}\right) = \frac{33(5)+33(12i)-4i(5)-4i(12i)}{5^2+12^2} = \frac{165+396i-20i-48i^2}{25+144}$

$= \frac{165+376i-48(-1)}{169} = \frac{165+376i+48}{169} = \frac{213+376i}{169} = \frac{213}{169} + \frac{376}{169}i$

Check the answer: $\frac{213+376i}{169}(3-2i)^2 = \frac{213+376i}{169}(3-2i)(3-2i)$

$= \frac{213+376i}{169}[3(3)+3(-2i)-2i(3)-2i(-2i)] = \frac{213+376i}{169}(9-6i-6i+4i^2)$

$= \frac{213+376i}{169}(9-12i-4) = \frac{213+376i}{169}(5-12i) = \frac{213(5)+213(-12i)+376i(5)+376i(-12i)}{169}$

$= \frac{1065-2556i+1880i-4512i^2}{169} = \frac{1065-676i+4512}{169} = \frac{5577-676i}{169} = \frac{5577}{169} - \frac{676i}{169} = 33-4i$

agrees with $(8+i)(4-i) = 8(4)+8(-i)+i(4)+i(-i) = 32-8i+4i-i^2$

$32-4i-(-1) = 32-4i+1 = 33-4i$

Chapter 5

(1) A: $6 + 4i$, B: $7 - 7i$, C: $-8 - 3i$, D: $8i$, E: $-4 + 5i$, F: $8 + 10i$, G: -7, H: $-3 - 9i$, I: $1 - 6i$, J: $-9 + 7i$ (Quadrant I: A & F, Q II: E & J, Q III: C & H, Q IV: B & I)

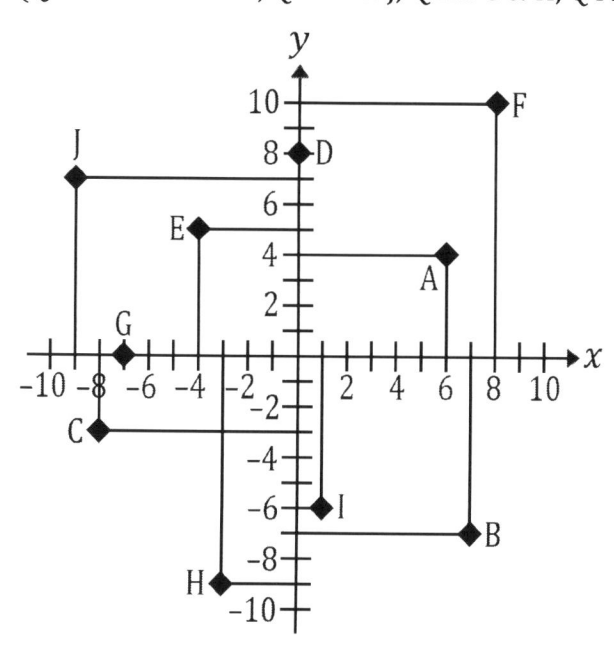

(2) $z = -8 + 3i$ (Quad. II), $\bar{z} = -8 - 3i$ (Q III), $-z = 8 - 3i$ (Q IV), $-\bar{z} = 8 + 3i$ (Q I)

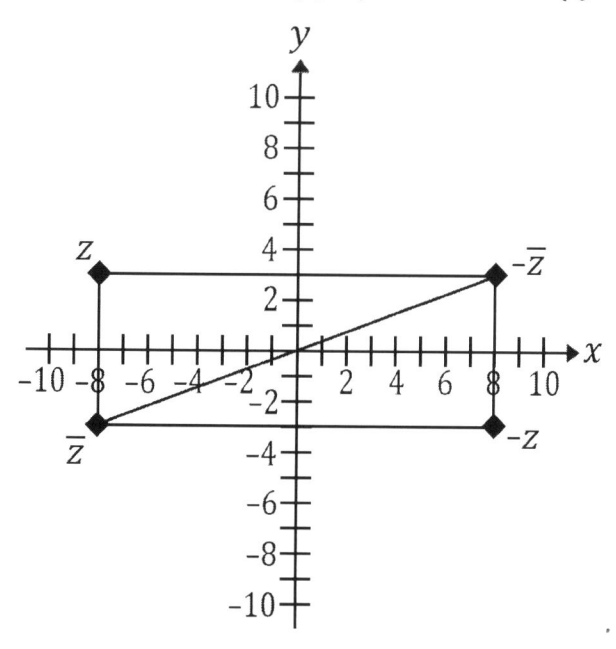

Answer Key

Chapter 6

(1) $|a| = |12 + 5i| = \sqrt{12^2 + 5^2} = \sqrt{144 + 25} = \sqrt{169} = 13$

(2) $|b| = |8 - 15i| = \sqrt{8^2 + (-15)^2} = \sqrt{64 + 225} = \sqrt{289} = 17$

(3) $|a + b| = |12 + 5i + 8 - 15i| = |(12 + 8) + (5i - 15i)| = |20 - 10i| = \sqrt{20^2 + (-10)^2} = \sqrt{400 + 100} = \sqrt{500} = \sqrt{100}\sqrt{5} = 10\sqrt{5}$

Notes: First add the complex numbers like we did in Chapter 4, and then use the formula for the modulus. In the last steps, we used the rule $\sqrt{pq} = \sqrt{p}\sqrt{q}$ to factor out the perfect square 100. Many instructors prefer the answer $10\sqrt{5}$ instead of $\sqrt{500}$.

(4) $|a - b| = |12 + 5i - (8 - 15i)| = |12 + 5i - 8 - (-15i)| = |12 + 5i - 8 + 15i|$
$= |(12 - 8) + (5i + 15i)| = |4 + 20i| = \sqrt{4^2 + 20^2} = \sqrt{16 + 400} = \sqrt{416} = \sqrt{16}\sqrt{26} = 4\sqrt{26}$

Note: In the last steps, we factored out the perfect square 16; note that $16 \times 26 = 416$. Many instructors prefer the answer $4\sqrt{26}$ instead of $\sqrt{416}$.

(5) $|3a - 2b| = |3(12 + 5i) - 2(8 - 15i)| = |36 + 15i - 16 - (-30i)| = |36 + 15i - 16 + 30i| = |(36 - 16) + (15i + 30i)| = |20 + 45i| = \sqrt{20^2 + 45^2} = \sqrt{400 + 2025} = \sqrt{2425} = \sqrt{25}\sqrt{97} = 5\sqrt{97}$

Note: In the last steps, we factored out the perfect square 100. Many instructors prefer the answer $5\sqrt{97}$ instead of $\sqrt{2425}$.

(6) $|ab| = |(12 + 5i)(8 - 15i)| = |12(8) + 12(-15i) + 5i(8) + 5i(-15i)| = |96 - 180i + 40i - 75i^2| = |96 - 140i - 75(-1)| = |96 - 140i + 75| = |171 - 140i| = \sqrt{171^2 + 140^2} = \sqrt{29{,}241 + 19{,}600} = \sqrt{48{,}841} = 221$

Alternate solution: $|ab| = |a||b| = (13)(17) = 221$ (see Chapter 16)

(7) $\left|\dfrac{a}{b}\right| = \left|\dfrac{12+5i}{8-15i}\right| = \left|\dfrac{12+5i}{8-15i}\left(\dfrac{8+15i}{8+15i}\right)\right| = \left|\dfrac{12(8)+12(15i)+5i(8)+5i(15i)}{8^2+15^2}\right| = \left|\dfrac{96+180i+40i+75i^2}{64+225}\right| = \left|\dfrac{96+220i+75(-1)}{289}\right| = \left|\dfrac{21+220i}{289}\right| = \left|\dfrac{21}{289} + \dfrac{220}{289}i\right| = \sqrt{\left(\dfrac{21}{289}\right)^2 + \left(\dfrac{220}{289}\right)^2} = \sqrt{\dfrac{441}{83{,}521} + \dfrac{48{,}400}{83{,}521}} = \sqrt{\dfrac{48{,}841}{83{,}521}} = \dfrac{221}{289} = \dfrac{(13)(17)}{(17)(17)} = \dfrac{13}{17}$

Alternate solution: $\left|\dfrac{a}{b}\right| = \dfrac{|a|}{|b|} = \dfrac{13}{17}$ (see Chapter 16)

Notes: Recall the method from Chapter 4 for dividing complex numbers. We reduced $\frac{221}{289}$ to $\frac{13}{17}$ using the fact that 221 and 289 are each evenly divisible by 17.

(8) $|a|^2 = |12 + 5i|^2 = \left(\sqrt{12^2 + 5^2}\right)^2 = 12^2 + 5^2 = 144 + 25 = 169$

(9) $|a^2| = |(12 + 5i)(12 + 5i)| = |12(12) + 12(5i) + 5i(12) + 5i(5i)| =$
$|144 + 60i + 60i + 25i^2| = |144 + 120i + 25(-1)| = |119 + 120i| =$
$\sqrt{119^2 + 120^2} = \sqrt{14,161 + 14,400} = \sqrt{28,561} = 169$

Notes: We see that $|a|^2 = |a^2|$. In Exercises 6-7, we saw that $|ab| = |a||b|$ and $\left|\frac{a}{b}\right| = \frac{|a|}{|b|}$. Challenge: Prove whether or not these relationships hold in general. (Addition and subtraction are different; Exercises 3-4 clearly show that $|a \pm b|$ **aren't** equal to $|a| \pm |b|$.) Chapter 16 will explore these and other identities.

(10) $|t| = |2 - i| = \sqrt{2^2 + (-1)^2} = \sqrt{4 + 1} = \sqrt{5}$ (see Example 2)

(11) $|u| = |2 + i\sqrt{3}| = \sqrt{2^2 + \left(\sqrt{3}\right)^2} = \sqrt{4 + 3} = \sqrt{7}$

Note: Recall the rule from algebra that $\left(\sqrt{a}\right)^2 = \sqrt{a}\sqrt{a} = a$. For example, $\left(\sqrt{3}\right)^2 = \sqrt{3}\sqrt{3} = \sqrt{9} = 3$.

(12) $|w| = \left|\frac{2}{i}\right| = \left|\frac{2}{i}\frac{i}{i}\right| = \left|\frac{2i}{i^2}\right| = \left|\frac{2i}{-1}\right| = |-2i| = \sqrt{0^2 + (-2)^2} = \sqrt{4} = 2$

Alternate solution: $|w| = \left|\frac{2}{i}\right| = \frac{|2|}{|i|} = \frac{2}{1} = 2$

(13) First find $w = \frac{2}{i} = \frac{2}{i}\frac{i}{i} = \frac{2i}{i^2} = \frac{2i}{-1} = -2i$ and $\overline{w} = 2i$ (recall Chapter 3).
$|t + \overline{w}| = |2 - i + 2i| = |2 + i| = \sqrt{2^2 + 1^2} = \sqrt{4 + 1} = \sqrt{5}$ (see Example 2)

(14) First find $w = \frac{2}{i} = \frac{2}{i}\frac{i}{i} = \frac{2i}{i^2} = \frac{2i}{-1} = -2i$.
$|u - w\sqrt{3}| = |2 + i\sqrt{3} - (-2i)\sqrt{3}| = |2 + i\sqrt{3} + 2i\sqrt{3}| = |2 + 3i\sqrt{3}| =$
$\sqrt{2^2 + \left(3\sqrt{3}\right)^2} = \sqrt{4 + 3^2\left(\sqrt{3}\right)^2} = \sqrt{4 + (9)(3)} = \sqrt{4 + 27} = \sqrt{31}$

Notes: Combine like terms to see that $i\sqrt{3} + 2i\sqrt{3} = 3i\sqrt{3}$; it's no different than $ab + 2ab = 3ab$. Note that $\left(3\sqrt{3}\right)^2 = 3^2\left(\sqrt{3}\right)^2$ according to $(ab)^c = a^c b^c$ and that $\left(\sqrt{3}\right)^2 = \sqrt{3}\sqrt{3} = \sqrt{9} = 3$.

Answer Key

(15) First find $\bar{t} = \overline{2-i} = 2+i, \bar{u} = \overline{2+i\sqrt{3}} = 2-i\sqrt{3}$, and $\bar{w} = 2i$ (as shown in the solution to Exercise 13).

$|\bar{t} + \bar{u}\sqrt{3} + \bar{w}| = |2+i+(2-i\sqrt{3})\sqrt{3}+2i| = |2+i+2\sqrt{3}-3i+2i| =$
$|(2+2\sqrt{3})+(i-3i+2i)| = |(2+2\sqrt{3})+(-3i+3i)| = |(2+2\sqrt{3})+0i| =$
$|(2+2\sqrt{3})| = \sqrt{(2+2\sqrt{3})^2} = 2+2\sqrt{3}$

Notes: $\sqrt{(2+2\sqrt{3})^2} = 2+2\sqrt{3}$ according to the rule $(\sqrt{a})^2 = \sqrt{a}\sqrt{a} = a$ (see the solution to Exercise 11). Combine like terms to see that $i - 3i + 2i = -3i + 3i = 0$. (The square root part is easy since the imaginary part of $\bar{t} + \bar{u}\sqrt{3} + \bar{w}$ canceled out.)

(16) First find $\bar{u} = \overline{2+i\sqrt{3}} = 2-i\sqrt{3}$.

$|t\bar{u}| = |(2-i)(2-i\sqrt{3})| = |2(2)+2(-i\sqrt{3})-i(2)-i(-i\sqrt{3})|$
$= |4-2i\sqrt{3}-2i+i^2\sqrt{3}| = |4-2i\sqrt{3}-2i-\sqrt{3}| = |(4-\sqrt{3})-(2\sqrt{3}+2)i| =$
$\sqrt{(4-\sqrt{3})^2 + (2\sqrt{3}+2)^2} = \sqrt{16-2(4)(\sqrt{3})+(\sqrt{3})^2+(2\sqrt{3})^2+2(2\sqrt{3})(2)+4}$
$= \sqrt{16-8\sqrt{3}+3+2^2(\sqrt{3})^2+8\sqrt{3}+4} = \sqrt{16+3+4(3)+4} = \sqrt{16+3+12+4}$
$= \sqrt{19+16} = \sqrt{35}$

Alternate solution: $|t\bar{u}| = |t||\bar{u}| = \sqrt{5}\sqrt{7}$ (using the answers to Exercises 10-11 along with $|\bar{u}| = |u| = \sqrt{7}$; see the note to the solution to Exercise 9).

Notes: We can work with $2\sqrt{3}+2$ instead of $-2\sqrt{3}-2$ since y gets squared in the formula $x^2 + y^2$. That is, $(-2\sqrt{3}-2)^2 = [(-1)(2\sqrt{3}+2)]^2 = (-1)^2(2\sqrt{3}+2)^2 = (1)(2\sqrt{3}+2)^2 = (2\sqrt{3}+2)^2$. Since $(a \pm b)^2 = a^2 \pm 2ab + b^2$, it follows that $(4-\sqrt{3})^2 = 16 - 2(4)(\sqrt{3})+(\sqrt{3})^2$ and $(2\sqrt{3}+2)^2 = (2\sqrt{3})^2 + 2(2\sqrt{3})(2) + 4$. It's really easy to make a mistake in the arithmetic, but if you enter

$\sqrt{(4-\sqrt{3})^2 + (2\sqrt{3}+2)^2}$ correctly on a calculator, you can verify that it agrees with the final answer of $\sqrt{35} \approx 5.916$. If you wish to check the calculator work, it is $\sqrt{2.267949^2 + 5.464102^2} \approx \sqrt{5.144 + 29.856} \approx \sqrt{35} \approx 5.916$.

(17) $\frac{|u|^2}{u} = \frac{(\sqrt{7})^2}{2+i\sqrt{3}} = \frac{7}{2+i\sqrt{3}}\left(\frac{2-i\sqrt{3}}{2-i\sqrt{3}}\right) = \frac{7(2-i\sqrt{3})}{2^2+(\sqrt{3})^2} = \frac{7(2-i\sqrt{3})}{4+3} = \frac{7(2-i\sqrt{3})}{7} = 2-i\sqrt{3}$

Alternate solution: Compare $|u| = \sqrt{x^2+y^2}$ with $u\bar{u} = (x+iy)(x-iy) = x^2+y^2$ (where $x=2$ and $y=\sqrt{3}$) to see that $|u|^2 = u\bar{u}$. Divide by u on both sides to get $\frac{|u|^2}{u} = \bar{u} = \overline{2+i\sqrt{3}} = 2-i\sqrt{3}$.

Note: Recall from Exercise 11 that $|u| = \sqrt{7}$. Recall that Chapter 4 showed how to divide by a complex number.

(18) $\frac{|u|^2}{t} = \frac{(\sqrt{7})^2}{2-i} = \frac{7}{2-i}\left(\frac{2+i}{2+i}\right) = \frac{14+7i}{2^2+1^2} = \frac{14+7i}{4+1} = \frac{14+7i}{5} = \frac{14}{5} + \frac{7}{5}i$

Note: Recall from Exercise 11 that $|u| = \sqrt{7}$. Recall that Chapter 4 showed how to divide by a complex number.

Answer Key

Chapter 7

(1) $|z| = \sqrt{x^2 + y^2} = \sqrt{10^2 + 24^2} = \sqrt{100 + 576} = \sqrt{676} = 26$

$\theta = \tan^{-1}\left(\frac{y}{x}\right) = \tan^{-1}\left(\frac{24}{10}\right) = \tan^{-1}(2.4) \approx 67.38°$ Notes: $x > 0, y > 0$ (Quad. I)

(2) $|z| = \sqrt{x^2 + y^2} = \sqrt{7^2 + (-7)^2} = \sqrt{49 + 49} = \sqrt{(2)(49)} = 7\sqrt{2}$

$\theta = \tan^{-1}\left(\frac{y}{x}\right) = \tan^{-1}\left(\frac{-7}{7}\right) = \tan^{-1}(-1) = -45°$ Alternate answer: 315°

Notes: $x > 0, y < 0$ (Quad. IV). Factor out the perfect square 49 to see that $\sqrt{98} = \sqrt{(49)(2)} = \sqrt{49}\sqrt{2} = 7\sqrt{2}$. (You can check that $\sqrt{98} \approx 9.899$ and $7\sqrt{2} \approx 9.899$.)

(3) $|z| = \sqrt{x^2 + y^2} = \sqrt{\left(-\sqrt{3}\right)^2 + 1^2} = \sqrt{3 + 1} = \sqrt{4} = 2$

$\theta = \tan^{-1}\left(\frac{y}{x}\right) = \tan^{-1}\left(\frac{1}{-\sqrt{3}}\right) = -30° + 180° = 150°$

Notes: $x < 0, y > 0$ (Quad. II), add 180° to the calculator's $\tan^{-1}$ since $x < 0$. Identify $x = -\sqrt{3}$ and $y = 1$ (the coefficient of i).

(4) $|z| = \sqrt{x^2 + y^2} = \sqrt{0^2 + 5^2} = \sqrt{25} = 5$, $\theta = \tan^{-1}\left(\frac{y}{x}\right) = \tan^{-1}\left(\frac{5}{0}\right) = 90°$

Notes: $x = 0, y > 0$ (on the imaginary axis). Although $\frac{5}{0}$ is undefined, we can still determine that $\tan^{-1}\left(\frac{5}{0}\right) = 90°$. One way is to realize that the point $(0, 5i)$ lies on the $+y$-axis, which is perpendicular to the x-axis. (The axes lie at 0°, 90°, 180°, and 270°. If a point lies on an axis, it helps to know these angles.) Another way is to approach $\tan^{-1}\left(\frac{5}{0}\right)$ as a limit. You can check that $\tan^{-1}\left(\frac{5}{0.1}\right) \approx 88.85°$, $\tan^{-1}\left(\frac{5}{0.01}\right) \approx 89.89°$, $\tan^{-1}\left(\frac{5}{0.001}\right) \approx 89.99°$, etc. The closer the denominator gets to zero, the closer the angle gets to 90°. (The idea of approaching the problem as a limit is one of the fundamental approaches to the subject of calculus.)

(5) $|z| = \sqrt{x^2 + y^2} = \sqrt{\left(-\sqrt{2}\right)^2 + \left(-\sqrt{6}\right)^2} = \sqrt{2 + 6} = \sqrt{8} = \sqrt{2(4)} = \sqrt{2}\sqrt{4} = 2\sqrt{2}$

$\theta = \tan^{-1}\left(\frac{y}{x}\right) = \tan^{-1}\left(\frac{-\sqrt{6}}{-\sqrt{2}}\right) = \tan^{-1}\left(\sqrt{3}\right) = 60° + 180° = 240°$

Notes: $x < 0, y < 0$ (Quad. III), add 180° to the calculator's $\tan^{-1}$ since $x < 0$. We factored out the perfect square 4 to get $\sqrt{8} = 2\sqrt{2}$. Also, $\frac{\sqrt{6}}{\sqrt{2}} = \frac{\sqrt{(3)(2)}}{\sqrt{2}} = \frac{\sqrt{3}\sqrt{2}}{\sqrt{2}} = \sqrt{3}$.

Complex Numbers Essentials Math Workbook with Answers

(6) First find $(4i)^4 = 4^4 i^4 = 256(1) = 256$, such that $x = 256$ and $y = 0$.
$|z| = \sqrt{x^2 + y^2} = \sqrt{256^2 + 0^2} = \sqrt{256^2} = 256$, $\theta = \tan^{-1}\left(\frac{y}{x}\right) = \tan^{-1} 0 = 0°$
Notes: $x > 0$, $y = 0$ (on the real axis).

(7) $|z| = \sqrt{x^2 + y^2} = \sqrt{2^2 + (-1)^2} = \sqrt{4+1} = \sqrt{5}$
$\theta = \tan^{-1}\left(\frac{y}{x}\right) = \tan^{-1}\left(\frac{-1}{2}\right) = \tan^{-1}(-0.5) \approx -26.57°$ (equivalent to $333.43°$)
Notes: $x > 0$, $y < 0$ (Quad. IV). Identify $x = 2$ and $y = -1$ (the coefficient of i).

(8) $|z| = \sqrt{x^2 + y^2} = \sqrt{(-11)^2 + 0^2} = \sqrt{121} = 11$, $\theta = \tan^{-1}\left(\frac{y}{x}\right) = \tan^{-1} 0 = 180°$
Notes: $x < 0$, $y = 0$ (on the negative real axis). The $-x$-axis lies at $180°$. (Since $x < 0$, add $180°$ to the calculator's answer for the inverse tangent.)

(9) First find $\frac{9}{i} = \frac{9i}{ii} = \frac{9i}{i^2} = \frac{9i}{-1} = -9i$ such that $x = 0$ and $y = -9$.
$|z| = \sqrt{x^2 + y^2} = \sqrt{0^2 + (-9)^2} = \sqrt{0 + 81} = 9$, $\theta = \tan^{-1}\left(\frac{y}{x}\right) = \tan^{-1}\left(\frac{-9}{0}\right) = 270°$
Alternate answer: $-90°$. Notes: $x = 0$, $y < 0$ (on the negative imaginary axis). Although $\frac{-9}{0}$ is undefined, we can still determine that $\tan^{-1}\left(\frac{-9}{0}\right) = -90°$ (or $270°$). One way is to realize that the point $(0, -9i)$ lies on the $-y$-axis, which lies at $270°$. Another way is to approach $\tan^{-1}\left(\frac{-9}{0}\right)$ as a limit. You can check that $\tan^{-1}\left(\frac{-9}{0.1}\right) \approx -89.36°$, $\tan^{-1}\left(\frac{-9}{0.01}\right) \approx -89.94°$, $\tan^{-1}\left(\frac{5}{0.001}\right) \approx -89.99°$, etc. The closer the denominator gets to zero, the closer the angle gets to $-90°$ (which equates to $270°$).

(10) $x = 8\cos 60° = 8\left(\frac{1}{2}\right) = 4$, $y = 8\sin 60° = 8\left(\frac{\sqrt{3}}{2}\right) = 4\sqrt{3}$, $z = 4 + 4i\sqrt{3}$
Note: The advantage of writing $4i\sqrt{3}$ instead of $4\sqrt{3}i$ is that it is easier to tell that the i isn't inside of the square root (especially when writing with your hand).

(11) $x = \sqrt{3}\cos 330° = \sqrt{3}\left(\frac{\sqrt{3}}{2}\right) = \frac{3}{2}$, $y = \sqrt{3}\sin 330° = \sqrt{3}\left(-\frac{1}{2}\right) = -\frac{\sqrt{3}}{2}$,
$z = \frac{3}{2} - \frac{\sqrt{3}}{2}i$ (Quadrant IV)

(12) $x = 13\cos 90° = 13(0) = 0$, $y = 13\sin 90° = 13(1) = 13$, $z = 13i$ (purely imag.)

(13) $x = \sqrt{2}\cos 225° = \sqrt{2}\left(-\frac{\sqrt{2}}{2}\right) = -\frac{2}{2} = -1$, $y = \sqrt{2}\sin 225° = \sqrt{2}\left(-\frac{\sqrt{2}}{2}\right) = -\frac{2}{2} = -1$, $z = -1 - i$ (Quadrant III)

Note: $\sqrt{2}\left(-\frac{\sqrt{2}}{2}\right) = -\frac{\sqrt{2}\sqrt{2}}{2} = -\frac{2}{2} = -1$

Answer Key

(14) $x = 17 \cos 128° \approx -10.47$, $y = 17 \sin 128° \approx 13.40$, $z \approx -10.47 + 13.4i$ (Quadrant II) Note: Check that your calculator is in **degrees** mode.

(15) $x = 4.8 \cos 196° \approx -4.614$, $y = 4.8 \sin 196° \approx -1.323$, $z = -4.614 - 1.323i$ (Quadrant III)

(16) $x = 25 \cos 304° \approx 13.98$, $y = 25 \sin 304° \approx -20.73$, $z = 13.98 - 20.73i$ (Quadrant IV)

(17) $x = 15 \cos 180° = 15(-1) = -15$, $y = 15 \sin 180° = 15(0) = 0$, $z = -15$ (on the negative real axis)

(18) $x = \sqrt{6} \cos 240° = \sqrt{6}\left(-\frac{1}{2}\right) = -\frac{\sqrt{6}}{2}$, $y = \sqrt{6} \sin 240° = \sqrt{6}\left(-\frac{\sqrt{3}}{2}\right) = -\frac{\sqrt{18}}{2} = -\frac{\sqrt{9(2)}}{2} = -\frac{3\sqrt{2}}{2}$, $z = -\frac{\sqrt{6}}{2} + -\frac{3i\sqrt{2}}{2}$ (Quadrant III)

Note: We factored out the perfect square 9 in $\frac{\sqrt{18}}{2} = \frac{\sqrt{9(2)}}{2} = \frac{\sqrt{9}\sqrt{2}}{2} = \frac{3\sqrt{2}}{2}$. (You can check that $\frac{\sqrt{18}}{2} \approx 2.121$ and $\frac{3\sqrt{2}}{2} \approx 2.121$.)

Chapter 8

(1) $\text{Re}(a)\text{Re}(b) = \text{Re}(4 - 3i)\text{Re}(12 + 5i) = (4)(12) = 48$

(2) First find $ab = (4 - 3i)(12 + 5i) = 4(12) + 4(5i) - 3i(12) - 3i(5i)$
$ab = 48 + 20i - 36i - 15i^2 = 48 - 16i - 15(-1) = 48 - 16i + 15 = 63 - 16i$
$\text{Re}(ab) = \text{Re}(63 - 16i) = 63$ Question: What can you conclude from Exercises 1-2?

(3) $5\text{Re}(b) - 10\text{Im}(a) = 5\text{Re}(12 + 5i) - 10\text{Im}(4 - 3i) = 5(12) - 10(-3) = 90$

(4) First find $c = \frac{2}{i} = \frac{2\,i}{i\,i} = \frac{2i}{i^2} = \frac{2i}{-1} = -2i$. Now find $\text{Im}(c) = \text{Im}(-2i) = -2$.

(5) First find $a + b + c = 4 - 3i + 12 + 5i - 2i = (4 + 12) + (-3i + 5i - 2i)$
$a + b + c = 48 + 0i = 48$. Now find $\text{Im}(48) = 0$ (since 48 is purely real).

Note: Recall from the solution to Exercise 4 that $c = \frac{2}{i} = -2i$

(6) $\frac{\text{Re}(a)\text{Im}(b)}{\text{Re}(b)+\text{Im}(a)} = \frac{\text{Re}(4-3i)\text{Im}(12+5i)}{\text{Re}(12+5i)+\text{Im}(4-3i)} = \frac{(4)(5)}{12+(-3)} = \frac{20}{12-3} = \frac{20}{9}$

(7) First find $\frac{b}{a} = \frac{12+5i}{4-3i} = \frac{12+5i}{4-3i}\left(\frac{4+3i}{4+3i}\right) = \frac{12(4)+12(3i)+5i(4)+5i(3i)}{4^2+3^2} = \frac{48+36i+20i+15i^2}{16+9}$
$\frac{b}{a} = \frac{48+56i+15(-1)}{25} = \frac{33+56i}{25} = \frac{33}{25} + \frac{56}{25}i$. Now find $\text{Im}\left(\frac{b}{a}\right) - \frac{\text{Im}(b)}{\text{Im}(a)} =$
$\text{Im}\left(\frac{33}{25} + \frac{56}{25}i\right) - \frac{\text{Im}(12+5i)}{\text{Im}(4-3i)} = \frac{56}{25} - \frac{5}{(-3)} = \frac{56}{25} + \frac{5}{3} = \frac{56}{25}\frac{3}{3} + \frac{5}{3}\frac{25}{25} = \frac{168}{75} + \frac{125}{75} = \frac{293}{75}$

Note: To add $\frac{56}{25} + \frac{5}{3}$, we made a common denominator of 75.

(8) $\frac{b+\bar{b}}{2} - \text{Re}(b) = \frac{12+5i+12-5i}{2} - \text{Re}(12+5i) = \frac{24}{2} - 12 = 12 - 12 = 0$

Note: $\bar{b}$ is the complex conjugate of b (Chapter 3).

(9) $\frac{b-\bar{b}}{2i} + \text{Im}(b) = \frac{12+5i-(12-5i)}{2i} + \text{Im}(12+5i) = \frac{12+5i-12-(-5i)}{2i} + 5 = \frac{5i+5i}{2i} + 5 =$
$\frac{10i}{2i} + 5 = 5 + 5 = 10$

Challenge: Prove that $\frac{z+\bar{z}}{2} = \text{Re}(z)$ and $\frac{z-\bar{z}}{2i} = \text{Im}(z)$ hold for any number z. (We will explore a variety of properties of complex numbers in Chapter 16.)

Answer Key

Chapter 9

(1) $5z^2 = 80 \to z^2 = \frac{80}{5} \to z^2 = 16 \to z = \pm\sqrt{16} \to z = \pm 4$

Check: $5z^2 = 5(\pm 4)^2 = 5(16) = 80$

(2) $3z^2 + 12 = 0 \to 3z^2 = -12 \to z^2 = -\frac{12}{3} \to z^2 = -4 \to z = \pm\sqrt{-4} \to$
$z = \pm\sqrt{-1}\sqrt{4} \to z = \pm 2i$

Check: $3z^2 + 12 = 3(\pm 2i)^2 + 12 = 3(\pm 2)^2 i^2 + 12 = 3(4)(-1) + 12 = -12 + 12 = 0$

(3) $6z^2 - 2 = 0 \to 6z^2 = 2 \to z^2 = \frac{2}{6} \to z^2 = \frac{1}{3} \to z = \pm\frac{1}{\sqrt{3}} \to z = \pm\frac{1}{\sqrt{3}}\frac{\sqrt{3}}{\sqrt{3}} = \pm\frac{\sqrt{3}}{3}$

Check: $6z^2 - 2 = 6\left(\pm\frac{\sqrt{3}}{3}\right)^2 - 2 = 6\frac{(\pm\sqrt{3})^2}{3^2} - 2 = \frac{6(3)}{9} - 2 = \frac{18}{9} - 2 = 2 - 2 = 0$

Note: We multiplied by $\frac{\sqrt{3}}{\sqrt{3}}$ in order to rationalize the denominator. The answer $\frac{\sqrt{3}}{3}$ is equivalent to the answer $\frac{1}{\sqrt{3}}$, but many instructors prefer the form $\frac{\sqrt{3}}{3}$.

(4) $8z^2 = -18 \to z^2 = -\frac{18}{8} \to z^2 = -\frac{9}{4} \to z = \pm\sqrt{-\frac{9}{4}} \to z = \pm\frac{\sqrt{-1}\sqrt{9}}{\sqrt{4}} = \pm\frac{3i}{2}$

Check: $8z^2 = 8\left(\pm\frac{3i}{2}\right)^2 = 8\frac{(\pm 3)^2 i^2}{2^2} = \frac{8(9)(-1)}{4} = -\frac{72}{4} = -18$

(5) First put the equation in standard form: $z^2 - 16z + 48 = 0$

$(z - 4)(z - 12) = 0 \to z = 4$ or $z = 12$

Two real roots since $b^2 - 4ac = (-16)^2 - 4(1)(48) = 256 - 192 = 64 > 0$

Alternate solution: $a = 1$, $b = -16$, and $c = 48$ (from $z^2 - 16z + 48 = 0$)

$z = \frac{-b \pm \sqrt{b^2 - 4ac}}{2a} = \frac{-(-16) \pm \sqrt{(-16)^2 - 4(1)(48)}}{2(1)} = \frac{16 \pm \sqrt{256 - 192}}{2}$

$z = \frac{16 \pm \sqrt{64}}{2} = \frac{16 \pm 8}{2} = 8 \pm 4 \to z = 8 - 4 = 4$ or $z = 8 + 4 = 12$

Check: $z^2 + 48 = 4^2 + 48 = 16 + 48 = 64$ agrees with $16z = 16(4) = 64$ and
$z^2 + 48 = 12^2 + 48 = 144 + 48 = 192$ agrees with $16z = 16(12) = 192$.

(6) First put the equation in standard form: $6z^2 + z - 2 = 0$

$(2z - 1)(3z + 2) = 0 \to 2z = 1$ or $3z = -2 \to z = \frac{1}{2}$ or $z = -\frac{2}{3}$

Two real roots since $b^2 - 4ac = 1^2 - 4(6)(-2) = 1 + 48 = 49 > 0$

Alternate solution: $a = 6$, $b = 1$, and $c = -2$ (from $6z^2 + z - 2 = 0$)

$$z = \frac{-b \pm \sqrt{b^2 - 4ac}}{2a} = \frac{-1 \pm \sqrt{1^2 - 4(6)(-2)}}{2(6)} = \frac{-1 \pm \sqrt{1 + 48}}{12}$$

$$z = \frac{-1 \pm \sqrt{49}}{12} = \frac{-1 \pm 7}{12} \to z = \frac{-1 - 7}{12} = -\frac{8}{12} = -\frac{2}{3} \text{ or } z = \frac{-1 + 7}{12} = \frac{6}{12} = \frac{1}{2}$$

Check: $6z^2 + z = 6\left(-\frac{2}{3}\right)^2 + \left(-\frac{2}{3}\right) = \frac{6(-2)^2}{3^2} - \frac{2}{3} = \frac{6(4)}{9} - \frac{2}{3} = \frac{24}{9} - \frac{6}{9} = \frac{18}{9} = 2$ and

$6z^2 + z = 6\left(\frac{1}{2}\right)^2 + \frac{1}{2} = \frac{6}{4} + \frac{2}{4} = \frac{8}{4} = 2$.

(7) First put the equation in standard form: $10z^2 - 9z - 40 = 0$

$(2z - 5)(5z + 8) = 0 \to 2z = 5 \text{ or } 5z = -8 \to z = \frac{5}{2} \text{ or } z = -\frac{8}{5}$

Two real roots since $b^2 - 4ac = (-9)^2 - 4(10)(-40) = 81 + 1600 = 1681 > 0$

Alternate solution: $a = 10$, $b = -9$, and $c = -40$ (from $10z^2 - 9z - 40 = 0$)

$$z = \frac{-b \pm \sqrt{b^2 - 4ac}}{2a} = \frac{-(-9) \pm \sqrt{(-9)^2 - 4(10)(-40)}}{2(10)} = \frac{9 \pm \sqrt{81 + 1600}}{20}$$

$$z = \frac{9 \pm \sqrt{1681}}{20} = \frac{9 \pm 41}{20} \to z = \frac{9 - 41}{20} = \frac{-32}{20} = -\frac{8}{5} \text{ or } z = \frac{9 + 41}{20} = \frac{50}{20} = \frac{5}{2}$$

Check: $10z^2 = 10\left(-\frac{8}{5}\right)^2 = 25.6$ agrees with $9z + 40 = 9\left(-\frac{8}{5}\right) + 40 = 25.6$ and

$10z^2 = 10\left(\frac{5}{2}\right)^2 = 62.5$ agrees with $9z + 40 = 9\left(\frac{5}{2}\right) + 40 = 62.5$.

(8) This equation is already in standard form: $4z^2 + 12z + 9 = 0$

$(2z + 3)(2z + 3) = 0 \to z = -\frac{3}{2}$ (a double root; both roots are the same)

A double real root since $b^2 - 4ac = 12^2 - 4(4)(9) = 144 - 144 = 0$

Alternate solution: $a = 4$, $b = 12$, and $c = 9$ (from $4z^2 + 12z + 9 = 0$)

$$z = \frac{-b \pm \sqrt{b^2 - 4ac}}{2a} = \frac{-12 \pm \sqrt{12^2 - 4(4)(9)}}{2(4)} = \frac{-12 \pm \sqrt{144 - 144}}{8} = \frac{-12}{8} = -\frac{3}{2}$$

Check: $4z^2 + 12z + 9 = 4\left(-\frac{3}{2}\right)^2 + 12\left(-\frac{3}{2}\right) + 9 = \frac{4(9)}{4} - \frac{36}{2} + 9 = 9 - 18 + 9 = 0$.

(9) First put the equation in standard form: $z^2 + z\sqrt{2} - 4 = 0$

$(z - \sqrt{2})(z + 2\sqrt{2}) = 0 \to z = \sqrt{2}$ or $z = -2\sqrt{2}$ (this one isn't as easy to factor)

Two real roots since $b^2 - 4ac = (\sqrt{2})^2 - 4(1)(-4) = 2 + 16 = 18 > 0$

Alternate solution: $a = 1$, $b = \sqrt{2}$, and $c = -4$ (from $z^2 + z\sqrt{2} - 4 = 0$)

Answer Key

$$z = \frac{-b \pm \sqrt{b^2 - 4ac}}{2a} = \frac{-\sqrt{2} \pm \sqrt{(\sqrt{2})^2 - 4(1)(-4)}}{2(1)} = \frac{-\sqrt{2} \pm \sqrt{2+16}}{2}$$

$$z = \frac{-\sqrt{2} \pm \sqrt{18}}{2} = \frac{-\sqrt{2} \pm \sqrt{(9)(2)}}{2} = \frac{-\sqrt{2} \pm 3\sqrt{2}}{2}$$

$$z = \frac{-\sqrt{2} - 3\sqrt{2}}{2} = \frac{-4\sqrt{2}}{2} = -2\sqrt{2} \text{ or } z = \frac{-\sqrt{2} + 3\sqrt{2}}{2} = \frac{2\sqrt{2}}{2} = \sqrt{2}$$

Notes: We factored out the perfect square 9 in $\sqrt{18} = \sqrt{(9)(2)} = \sqrt{9}\sqrt{2} = 3\sqrt{2}$. We combined like terms in $-\sqrt{2} - 3\sqrt{2} = -4\sqrt{2}$ and $-\sqrt{2} + 3\sqrt{2} = 2\sqrt{2}$.

Check: $z^2 + z\sqrt{2} = (\sqrt{2})^2 + \sqrt{2}\sqrt{2} = 2 + 2 = 4$ and $z^2 + z\sqrt{2} = (-2\sqrt{2})^2 + (-2\sqrt{2})\sqrt{2} = (-2)^2(\sqrt{2})^2 - 2(\sqrt{2})^2 = 4(2) - 2(2) = 8 - 4 = 4$.

(10) First put the equation in standard form: $z^2 + 8iz - 15 = 0$

$(z + 3i)(z + 5i) = 0 \to z = -3i$ or $z = -5i$ (note that $i^2 = -1$)

(Since $b = 8i$ isn't real, the usual discriminant rules don't apply.)

Alternate solution: $a = 1$, $b = 8i$, and $c = -15$ (from $z^2 + 8iz - 15 = 0$)

$$z = \frac{-b \pm \sqrt{b^2 - 4ac}}{2a} = \frac{-8i \pm \sqrt{(8i)^2 - 4(1)(-15)}}{2(1)} = \frac{-8i \pm \sqrt{-64+60}}{2}$$

$$z = \frac{-8i \pm \sqrt{-4}}{2} = \frac{-8i \pm \sqrt{(4)(-1)}}{2} = \frac{-8i \pm \sqrt{4}\sqrt{-1}}{2} = \frac{-8i \pm 2i}{2} = -4i \pm i$$

$$z = -4i - i = -5i \text{ or } z = -4i + i = -3i$$

Check: $z^2 + 8iz = (-5i)^2 + 8i(-5i) = (-5)^2 i^2 - 40i^2 = 25(-1) - 40(-1) = -25 + 40 = 15$ and $z^2 + 8iz = (-3i)^2 + 8i(-3i) = (-3)^2 i^2 - 24i^2 = 9(-1) - 24(-1) = -9 + 24 = 15$.

(11) $(3z - i)(4z + 9i) = 0 \to 3z = i$ or $4z = -9i \to z = \frac{i}{3}$ or $z = -\frac{9}{4}i$

(Since $b = 23i$ isn't real, the usual discriminant rules don't apply.)

Alternate solution: $a = 12$, $b = 23i$, and $c = 9$ (from $12z^2 + 23iz + 9 = 0$)

$$z = \frac{-b \pm \sqrt{b^2 - 4ac}}{2a} = \frac{-23i \pm \sqrt{(23i)^2 - 4(12)(9)}}{2(12)} = \frac{-23i \pm \sqrt{-529-432}}{24}$$

$$z = \frac{-23i \pm \sqrt{-961}}{24} = \frac{-23i \pm \sqrt{(961)(-1)}}{24} = \frac{-23i \pm \sqrt{961}\sqrt{-1}}{24} = \frac{-23i \pm 31i}{24}$$

$$z = \frac{-23i - 31i}{24} = \frac{-54i}{24} = -\frac{9}{4}i \text{ or } z = \frac{-23i + 31i}{24} = \frac{8i}{24} = \frac{i}{3}$$

Check: $12z^2 + 23iz + 9 = 12\left(-\frac{9}{4}i\right)^2 + 23i\left(-\frac{9}{4}i\right) + 9 = 12\frac{(-9)^2 i^2}{4^2} + 23\frac{(-9)i^2}{4} + 9 =$
$\frac{12(81)(-1)}{16} + \frac{23(-9)(-1)}{4} + 9 = -\frac{972}{16} + \frac{207}{4} + 9 = -\frac{972}{16} + \frac{828}{16} + \frac{144}{16} = 0$ and $12z^2 +$
$23iz + 9 = 12\left(\frac{i}{3}\right)^2 + 23i\left(\frac{i}{3}\right) + 9 = 12\frac{(-1)}{9} + 23\frac{(-1)}{3} + 9 = -\frac{4}{3} - \frac{23}{3} + \frac{27}{3} = 0$.

(12) First put the equation in standard form: $z^2 - 4z - 9iz + 36i = 0$
This solution is **much simpler** if you are able to **factor** it as follows:
$(z - 4)(z - 9i) = 0 \rightarrow z = 4$ or $z = 9i$
(Since $b = -4 - 9i$ and $c = 36i$ aren't real, the usual discriminant rules don't apply.)
Alternate solution: The quadratic formula is more challenging in this problem:
$a = 1$, $b = -4 - 9i$, and $c = 36i$ (from $z^2 - 4z - 9iz + 36i = 0$)

$$z = \frac{-b \pm \sqrt{b^2 - 4ac}}{2a} = \frac{-(-4 - 9i) \pm \sqrt{(-4 - 9i)^2 - 4(1)(36i)}}{2(1)}$$

$$z = \frac{4 + 9i \pm \sqrt{(-4 - 9i)^2 - 144i}}{2}$$

$$z = \frac{4 + 9i \pm \sqrt{(-4)^2 + (-4)(-9i) + (-9i)(-4) + (-9i)^2 - 144i}}{2}$$

$$z = \frac{4 + 9i \pm \sqrt{16 + 36i + 36i + 81i^2 - 144i}}{2}$$

$$z = \frac{4 + 9i \pm \sqrt{16 + 72i - 81 - 144i}}{2} = \frac{4 + 9i \pm \sqrt{-65 - 72i}}{2}$$

The **trick** here is to ask yourself which complex number squared happens to equal $-65 - 72i$. Don't worry; you shouldn't 'know' this answer offhand, but you can figure it out. We want a complex number of the form $x + iy$ such that its square equals $-65 - 72i$. Write this in an equation:
$$(x + iy)^2 = -65 - 72i$$
Since $(x + iy)^2 = (x + iy)(x + iy) = x^2 + 2ixy - y^2$, this becomes:
$$x^2 + 2ixy - y^2 = -65 - 72i$$
For x and y to each be real, we need the real and imaginary parts of both sides of the above equation to be real. This gives us two equations:

Answer Key

$$x^2 - y^2 = -65 \quad \text{and} \quad 2ixy = -72i$$

Now we have two equations in two unknowns. We can solve this system by isolating y in the second equation: $y = -\frac{72i}{2ix} = -\frac{36}{x}$. Plug this into the first equation to get $x^2 - \left(-\frac{36}{x}\right)^2 = 79$, which simplifies to $x^2 - \frac{1296}{x^2} = -65$. Multiply by x^2 on both sides to get $x^4 - 1296 = -65x^2$. Make the substitution $u = x^2$ to get $u^2 - 1296 = -65u$. Put this in standard form: $u^2 + 65u - 1296 = 0$. Now we have a quadratic equation for the new variable u. Use the quadratic formula with $a = 1$, $b = 65$, and $c = -1296$.

$$u = \frac{-b \pm \sqrt{b^2 - 4ac}}{2a} = \frac{-65 \pm \sqrt{65^2 - 4(1)(-1296)}}{2(1)}$$

$$u = \frac{-65 \pm \sqrt{4225 + 5184}}{2} = \frac{-65 \pm \sqrt{9409}}{2} = \frac{-65 \pm 97}{2}$$

$$u = \frac{-65 + 97}{2} = \frac{32}{2} = 16 \text{ or } u = \frac{-65 - 97}{2} = \frac{-162}{2} = -81$$

We just need one solution for u, so let's pick $u = 16$. Since $u = x^2$, this gives $16 = x^2$. Since we only need one solution for x, let's pick the positive root: $x = 4$. This gives $y = -\frac{36}{x} = -\frac{36}{4} = -9$. Now we have the complex number $x + iy = 4 - 9i$. Now can you remember what we were doing? (If you think this was too much work, you're right. We'll use a **simpler method** to find $\sqrt{-65 - 72i}$ in Chapter 10, Example 2. It turns out to be easier to use trigonometry.) What we finally found is that $(4 - 9i)^2 = -65 - 72i$, which means that $\sqrt{-65 - 72i} = 4 - 9i$. This is easy to check: $(4 - 9i)(4 - 9i) = 16 - 36i - 36i + 81i^2 = -65 - 72i$. Now we may replace $\sqrt{-65 - 72i}$ with $4 - 9i$ back in the quadratic formula that we had been working out previously. We'll pick up where we left off.

$$z = \frac{4 + 9i \pm \sqrt{-65 - 72i}}{2} = \frac{4 + 9i \pm (4 - 9i)}{2}$$

$$z = \frac{4 + 9i - (4 - 9i)}{2} = \frac{4 + 9i - 4 + 9i}{2} = \frac{18i}{2} = 9i$$

$$\text{or } z = \frac{4 + 9i + (4 - 9i)}{2} = \frac{4 + 9i + 4 - 9i}{2} = \frac{8}{2} = 4$$

This is an extreme case of a problem that is **much, much easier** to solve by **factoring** than by using the quadratic formula. However, sometimes factoring is really difficult, so it pays to be fluent in both methods. As a counterexample, many students would have a tough time trying to factor Exercise 15, but would find the quadratic formula more straightforward.

Check: $z^2 + 36i = (9i)^2 + 36i = -81 + 36i$ agrees with $9iz + 4z = 9i(9i) + 4(9i) = 81i^2 + 36i = -81 + 36i$ and $z^2 + 36i = 4^2 + 36i = 16 + 36i$ agrees with $9iz + 4z = 9i(4) + 4(4) = 36i + 16$.

(13) First put the equation in standard form: $z^2 - 6z + 13 = 0$
$(z - 3 - 2i)(z - 3 + 2i) = 0 \to z = 3 + 2i$ or $z = 3 - 2i$ (not so easy to factor)
Two complex roots since $b^2 - 4ac = (-6)^2 - 4(1)(13) = 36 - 52 = -16 < 0$
Alternate solution: $a = 1$, $b = -6$, and $c = 13$ (from $z^2 - 6z + 13 = 0$)

$$z = \frac{-b \pm \sqrt{b^2 - 4ac}}{2a} = \frac{-(-6) \pm \sqrt{(-6)^2 - 4(1)(13)}}{2(1)} = \frac{6 \pm \sqrt{36 - 52}}{2}$$

$$z = \frac{6 \pm \sqrt{-16}}{2} = \frac{6 \pm \sqrt{16}\sqrt{-1}}{2} = \frac{6 \pm 4i}{2} = 3 \pm 2i \to z = 3 - 2i \text{ or } z = 3 + 2i$$

Check: $z^2 + 13 = (3 - 2i)(3 - 2i) + 13 = 9 - 6i - 6i + 4i^2 + 13 = 9 - 12i - 4 + 13 = 18 - 12i$ agrees with $6z = 6(3 - 2i) = 18 - 12i$ and $z^2 + 13 = (3 + 2i)(3 + 2i) + 13 = 9 + 6i + 6i + 4i^2 + 13 = 9 + 12i - 4 + 13 = 18 + 12i$ agrees with $6z = 6(3 + 2i) = 18 + 12i$.

(14) First put the equation in standard form: $z^2 - 8z + 11 = 0$
$(z - 4 - \sqrt{5})(z - 4 + \sqrt{5}) = 0 \to z = 4 + \sqrt{5}$ or $z = 4 - \sqrt{5}$ (not so easy to factor)
Two real roots since $b^2 - 4ac = (-8)^2 - 4(1)(11) = 64 - 44 = 20 > 0$
Alternate solution: $a = 1$, $b = -8$, and $c = 11$ (from $z^2 - 8z + 11 = 0$)

$$z = \frac{-b \pm \sqrt{b^2 - 4ac}}{2a} = \frac{-(-8) \pm \sqrt{(-8)^2 - 4(1)(11)}}{2(1)} = \frac{8 \pm \sqrt{64 - 44}}{2}$$

$$z = \frac{8 \pm \sqrt{20}}{2} = \frac{8 \pm \sqrt{(4)(5)}}{2} = \frac{8 \pm \sqrt{4}\sqrt{5}}{2} = \frac{8 \pm 2\sqrt{5}}{2} = 4 \pm \sqrt{5}$$

$$z = 4 - \sqrt{5} \text{ or } z = 4 + \sqrt{5}$$

Check: $z^2 + 11 = (4 - \sqrt{5})(4 - \sqrt{5}) + 11 = 16 - 4\sqrt{5} - 4\sqrt{5} + (\sqrt{5})^2 + 11 = 16 - 8\sqrt{5} + 5 + 11 = 32 - 8\sqrt{5}$ agrees with $8z = 8(4 - \sqrt{5}) = 32 - 8\sqrt{5}$ and $z^2 + 11 = (4 + \sqrt{5})(4 + \sqrt{5}) + 11 = 16 + 4\sqrt{5} + 4\sqrt{5} + (\sqrt{5})^2 + 11 = 16 + 8\sqrt{5} + 5 + 11 = 32 + 8\sqrt{5}$ agrees with $8z = 8(4 + \sqrt{5}) = 32 + 8\sqrt{5}$.

Answer Key

(15) First put the equation in standard form: $z^2 - 4z + 7 = 0$

$(z - 2 - i\sqrt{3})(z - 2 + i\sqrt{3}) = 0 \to z = 2 + i\sqrt{3}$ or $z = 2 - i\sqrt{3}$ (this one is difficult)

Two complex roots since $b^2 - 4ac = (-4)^2 - 4(1)(7) = 16 - 28 = -12 < 0$

Alternate solution: $a = 1$, $b = -4$, and $c = 7$ (from $z^2 - 4z + 7 = 0$)

$$z = \frac{-b \pm \sqrt{b^2 - 4ac}}{2a} = \frac{-(-4) \pm \sqrt{(-4)^2 - 4(1)(7)}}{2(1)} = \frac{4 \pm \sqrt{16 - 28}}{2}$$

$$z = \frac{4 \pm \sqrt{-12}}{2} = \frac{4 \pm \sqrt{(-1)(12)}}{2} = \frac{4 \pm \sqrt{-1}\sqrt{4}\sqrt{3}}{2} = \frac{4 \pm 2i\sqrt{3}}{2} = 2 \pm i\sqrt{3}$$

$$z = 2 - i\sqrt{3} \text{ or } z = 2 + i\sqrt{3}$$

Check: $z^2 + 7 = (2 - i\sqrt{3})(2 - i\sqrt{3}) + 7 = 4 - 2i\sqrt{3} - 2i\sqrt{3} + i^2(\sqrt{3})^2 + 7 = 4 - 4i\sqrt{3} - 3 + 7 = 8 - 4i\sqrt{3}$ agrees with $4z = 4(2 - i\sqrt{3}) = 8 - 4i\sqrt{3}$ and $z^2 + 7 = (2 + i\sqrt{3})(2 + i\sqrt{3}) + 7 = 4 + 2i\sqrt{3} + 2i\sqrt{3} + i^2(\sqrt{3})^2 + 7 = 4 + 4i\sqrt{3} - 3 + 7 = 8 + 4i\sqrt{3}$ agrees with $4z = 4(2 + i\sqrt{3}) = 8 + 4i\sqrt{3}$.

(16) First put the equation in standard form: $4z^2 - 20z + 34 = 0$

$(2z - 5 - 3i)(2z - 5 + 3i) = 0 \to z = \frac{5}{2} + \frac{3}{2}i$ or $z = \frac{5}{2} - \frac{3}{2}i$ (not so easy to factor)

Two complex roots since $b^2 - 4ac = (-20)^2 - 4(4)(34) = 400 - 544 = -144 < 0$

Alternate solution: $a = 4$, $b = -20$, and $c = 34$ (from $4z^2 - 20z + 34 = 0$)

$$z = \frac{-b \pm \sqrt{b^2 - 4ac}}{2a} = \frac{-(-20) \pm \sqrt{(-20)^2 - 4(4)(34)}}{2(4)} = \frac{20 \pm \sqrt{400 - 544}}{8}$$

$$z = \frac{20 \pm \sqrt{-144}}{8} = \frac{20 \pm \sqrt{144}\sqrt{-1}}{8} = \frac{20 \pm 12i}{8} = \frac{20}{8} \pm \frac{12i}{8} = \frac{5}{2} \pm \frac{3}{2}i$$

$$z = \frac{5}{2} - \frac{3}{2}i \text{ or } z = \frac{5}{2} + \frac{3}{2}i$$

Check: $4z^2 + 34 = 4\left(\frac{5}{2} \pm \frac{3}{2}i\right)\left(\frac{5}{2} \pm \frac{3}{2}i\right) + 34 = 4\left(\frac{25}{4} \pm \frac{15}{4}i \pm \frac{15}{4}i + \frac{9}{4}i^2\right) + 34 = 25 \pm 15i \pm 15i + 9i^2 + 34 = 25 \pm 30i - 9 + 34 = 50 \pm 30i$ agrees with $20z = 20\left(\frac{5}{2} \pm \frac{3}{2}i\right) = \frac{100}{2} \pm \frac{60}{2}i = 50 \pm 30i$.

Chapter 10

(1) Let $z = x + iy = 3 - 3i$, such that $x = 3$ and $y = -3$ (Quad. IV).

$$|z| = \sqrt{x^2 + y^2} = \sqrt{3^2 + (-3)^2} = \sqrt{9 + 9} = \sqrt{18} = \sqrt{(9)(2)} = \sqrt{9}\sqrt{2} = 3\sqrt{2}$$

$$\theta = \tan^{-1}\left(\frac{y}{x}\right) = \tan^{-1}\left(\frac{-3}{3}\right) = \tan^{-1}(-1) = -45° \text{ or } 315° \text{ (Quad. IV)}$$

$$(|z|\cos\theta + i|z|\sin\theta)^4 = |z|^4\cos(4\theta) + i|z|^4\sin(4\theta)$$

$$(3 - 3i)^4 = \left(3\sqrt{2}\right)^4\cos[4(-45°)] + \left(3\sqrt{2}\right)^4 i\sin[4(-45°)]$$

$$(3 - 3i)^4 = 3^4\left(\sqrt{2}\right)^4\cos(-180°) + 3^4\left(\sqrt{2}\right)^4 i\sin(-180°)$$

$$(3 - 3i)^4 = 81(4)(-1) + 81(4)i(0) = -324 + 0 = -324$$

Note: $\left(\sqrt{2}\right)^4 = \sqrt{2}\sqrt{2}\sqrt{2}\sqrt{2} = (2)(2) = 4$.

Check: $(3 - 3i)^2 = (3 - 3i)(3 - 3i) = 3(3) + 3(-3i) - 3i(3) - 3i(-3i)$
$= 9 - 9i - 9i + 9i^2 = 9 - 18i - 9 = -18i$
$(3 - 3i)^4 = (3 - 3i)^2(3 - 3i)^2 = (-18i)(-18i) = 324i^2 = -324$

(2) Let $z = x + iy = i - 2 = -2 + i$, such that $x = -2$ and $y = 1$ (Quad. II).

$$|z| = \sqrt{x^2 + y^2} = \sqrt{(-2)^2 + 1^2} = \sqrt{4 + 1} = \sqrt{5}$$

$$\theta = \tan^{-1}\left(\frac{y}{x}\right) = \tan^{-1}\left(\frac{1}{-2}\right) \approx -26.565° + 180° \approx 153.435° \text{ (Quad. II)}$$

Note: Add $180°$ to the calculator's answer for inverse tangent if $x < 0$ (Chapter 7).

$$(|z|\cos\theta + i|z|\sin\theta)^6 = |z|^6\cos(6\theta) + i|z|^6\sin(6\theta)$$

$$(i - 2)^6 \approx \left(\sqrt{5}\right)^6\cos[6(153.435°)] + \left(\sqrt{5}\right)^6 i\sin[6(153.435°)]$$

$$(i - 2)^6 \approx 125\cos 920.61° + 125i\sin 920.61°$$

$$(3 - 3i)^6 = -117 - 44i$$

Note: $\left(\sqrt{5}\right)^6 = \sqrt{5}\sqrt{5}\sqrt{5}\sqrt{5}\sqrt{5}\sqrt{5} = (5)(5)(5) = 125$.

Check: $(i - 2)^2 = (i - 2)(i - 2) = i^2 + i(-2) - 2i - 2(-2) = -1 - 2i - 2i + 4$
$= 3 - 4i$ such that $(i - 2)^4 = (i - 2)^2(i - 2)^2 = (3 - 4i)(3 - 4i)$
$= 3(3) + 3(-4i) - 4i(3) - 4i(-4i) = 9 - 12i - 12i + 16i^2 = 9 - 24i - 16$
$= -7 - 24i$ such that $(i - 2)^6 = (i - 2)^4(i - 2)^2 = (-7 - 24i)(3 - 4i)$
$= -7(3) - 7(-4i) - 24i(3) - 24i(-4i) = -21 + 28i - 72i + 96i^2$
$= -21 - 44i - 96 = -117 - 44i$

Answer Key

(3) Let $z = x + iy = \sqrt{2} + i\sqrt{6}$, such that $x = \sqrt{2}$ and $y = \sqrt{6}$ (Quad. I).

$$|z| = \sqrt{x^2 + y^2} = \sqrt{(\sqrt{2})^2 + (\sqrt{6})^2} = \sqrt{2 + 6} = \sqrt{8} = \sqrt{(4)(2)} = \sqrt{4}\sqrt{2} = 2\sqrt{2}$$

$$\theta = \tan^{-1}\left(\frac{y}{x}\right) = \tan^{-1}\left(\frac{\sqrt{6}}{\sqrt{2}}\right) = \tan^{-1}\left(\sqrt{\frac{6}{2}}\right) = \tan^{-1}(\sqrt{3}) = 60° \text{ (Quad. I)}$$

$$(|z|\cos\theta + i|z|\sin\theta)^5 = |z|^5 \cos(5\theta) + i|z|^5 \sin(5\theta)$$

$$(\sqrt{2} + i\sqrt{6})^5 = (2\sqrt{2})^5 \cos[5(60°)] + (2\sqrt{2})^5 i\sin[5(60°)]$$

$$(\sqrt{2} + i\sqrt{6})^5 = 2^5(\sqrt{2})^5 \cos 300° + 2^5(\sqrt{2})^5 i\sin 300°$$

$$(\sqrt{2} + i\sqrt{6})^5 = 32(4\sqrt{2})\left(\frac{1}{2}\right) + 32(4\sqrt{2})i\left(-\frac{\sqrt{3}}{2}\right) = 64\sqrt{2} - 64i\sqrt{2}\sqrt{3} = 64\sqrt{2} - 64i\sqrt{6}$$

Note: $(\sqrt{2})^5 = (\sqrt{2}\sqrt{2})(\sqrt{2}\sqrt{2})\sqrt{2} = (2)(2)\sqrt{2} = 4\sqrt{2}$.

Check: $(\sqrt{2} + i\sqrt{6})^2 = (\sqrt{2} + i\sqrt{6})(\sqrt{2} + i\sqrt{6}) = \sqrt{2}\sqrt{2} + i\sqrt{2}\sqrt{6} + i\sqrt{6}\sqrt{2} + i^2\sqrt{6}\sqrt{6}$

$= 2 + i\sqrt{12} + i\sqrt{12} - 6 = -4 + 2i\sqrt{12} = -4 + 2i\sqrt{(4)(3)} = -4 + 4i\sqrt{3}$

$(\sqrt{2} + i\sqrt{6})^4 = (\sqrt{2} + i\sqrt{6})^2(\sqrt{2} + i\sqrt{6})^2 = (-4 + 4i\sqrt{3})(-4 + 4i\sqrt{3})$

$= -4(-4) - 4(4i\sqrt{3}) + 4i\sqrt{3}(-4) + 4i\sqrt{3}(4i\sqrt{3}) = 16 - 16i\sqrt{3} - 16i\sqrt{3} + 16i^2(3)$

$= 16 - 32i\sqrt{3} - 48 = -32 - 32i\sqrt{3}$

$(\sqrt{2} + i\sqrt{6})^5 = (\sqrt{2} + i\sqrt{6})^4(\sqrt{2} + i\sqrt{6}) = (-32 - 32i\sqrt{3})(\sqrt{2} + i\sqrt{6})$

$-32\sqrt{2} - 32(i\sqrt{6}) - 32i\sqrt{3}\sqrt{2} - 32i\sqrt{3}(i\sqrt{6}) = -32\sqrt{2} - 32i\sqrt{6} - 32i\sqrt{6} - 32i^2\sqrt{18}$

$= -32\sqrt{2} - 64i\sqrt{6} + 32\sqrt{(9)(2)} = -32\sqrt{2} - 64i\sqrt{6} + 96\sqrt{2} = 64\sqrt{2} - 64i\sqrt{6}$

(4) Let $z = x + iy = -3 - 4i$, such that $x = -3$ and $y = -4$ (Quad. III).

$$|z| = \sqrt{x^2 + y^2} = \sqrt{(-3)^2 + (-4)^2} = \sqrt{9 + 16} = \sqrt{25} = 5$$

$$\theta = \tan^{-1}\left(\frac{y}{x}\right) = \tan^{-1}\left(\frac{-4}{-3}\right) = \tan^{-1}\left(\frac{4}{3}\right) \approx 53.1301° + 180° \approx 233.1301° \text{ (Quad. III)}$$

$$(|z|\cos\theta + i|z|\sin\theta)^7 = |z|^7\cos(7\theta) + i|z|^7\sin(7\theta)$$

$$(3 - 4i)^7 \approx 5^7 \cos[7(233.1301°)] + 5^7 i\sin[7(233.1301°)]$$

$$(3 - 4i)^7 \approx 78{,}125 \cos 1631.9107° + 78{,}125i \sin 1631.9107°$$

$$(3 - 4i)^7 = -76{,}443 - 16{,}124i$$

Check: $(-3 - 4i)^2 = [(-1)(3 + 4i)]^2 = (-1)^2(3 + 4i)(3 + 4i) = (3 + 4i)(3 + 4i)$

$$= 3(3) + 3(4i) + 4i(3) + 4i(4i) = 9 + 12i + 12i - 16 = -7 + 24i$$
$$(-3 - 4i)^4 = (-3 - 4i)^2(-3 - 4i)^2 = (-7 + 24i)(-7 + 24i)$$
$$= -7(-7) - 7(24i) + 24i(-7) + 24i(24i) = 49 - 168i - 168i + 576i^2$$
$$= 49 - 336i - 576 = -527 - 336i$$
$$(-3 - 4i)^6 = (-3 - 4i)^4(-3 - 4i)^2 = (-527 - 336i)(-7 + 24i)$$
$$= -527(-7) - 527(24i) - 336i(-7) - 336i(24i)$$
$$= 3689 - 12{,}648i + 2352i - 8064i^2 = 3689 - 10{,}296i + 8064 = 11{,}753 - 10{,}296i$$
$$(-3 - 4i)^7 = (-3 - 4i)^6(-3 - 4i) = (11{,}753 - 10{,}296i)(-3 - 4i)$$
$$= 11{,}753(-3) + 11{,}753(-4i) - 10{,}296i(-3) - 10{,}296i(-4i)$$
$$= -35{,}259 - 47{,}012i + 30{,}888i + 41{,}184i^2 = -35{,}259 - 16{,}124i - 41{,}184$$
$$= -76{,}443 - 16{,}124i$$

(5) Let $z = x + iy = i - \sqrt{2} = -\sqrt{2} + i$, such that $x = -\sqrt{2}$ and $y = 1$ (Quad. II).

$$|z| = \sqrt{x^2 + y^2} = \sqrt{\left(-\sqrt{2}\right)^2 + 1^2} = \sqrt{2 + 1} = \sqrt{3}$$

$$\theta = \tan^{-1}\left(\frac{y}{x}\right) = \tan^{-1}\left(\frac{1}{-\sqrt{2}}\right) \approx -35.2644 + 180° \approx 144.7356° \text{ (Quad. II)}$$

Note: Add 180° to the calculator's answer for inverse tangent if $x < 0$ (Chapter 7).

$$(|z|\cos\theta + i|z|\sin\theta)^8 = |z|^8 \cos(8\theta) + i|z|^8 \sin(8\theta)$$

Note: $\left(\sqrt{3}\right)^8 = \sqrt{3}\sqrt{3}\sqrt{3}\sqrt{3}\sqrt{3}\sqrt{3}\sqrt{3}\sqrt{3} = (3)(3)(3)(3) = 81$.

$$\left(i - \sqrt{2}\right)^8 \approx \left(\sqrt{3}\right)^8 \cos[8(144.7356°)] + \left(\sqrt{3}\right)^8 i\sin[8(144.7356°)]$$

$$\left(i - \sqrt{2}\right)^8 \approx 81\cos 1157.8848° + 81i\sin 1157.8848°$$

$$\left(i - \sqrt{2}\right)^8 \approx 17 + 79.1959i = 17 + 56i\sqrt{2}$$

Note: Although $81i\sin 1157.8848° \approx 79.1959i$, if you think about multiplying $\left(i - \sqrt{2}\right)^8$, it wouldn't be surprising for one of the numbers to include a $\sqrt{2}$. If you divide 79.1959 by $\sqrt{2}$, you get 56, showing that $79.1959 \approx 56\sqrt{2}$. The following check shows that 79.1959 is approximately the answer, whereas $56\sqrt{2}$ is the **exact** answer.

Check: $\left(i - \sqrt{2}\right)^2 = \left(i - \sqrt{2}\right)\left(i - \sqrt{2}\right) = i^2 + i\left(-\sqrt{2}\right) - \sqrt{2}(i) - \sqrt{2}\left(-\sqrt{2}\right)$

$$= -1 - i\sqrt{2} - i\sqrt{2} + 2 = 1 - 2i\sqrt{2} \text{ such that } \left(i - \sqrt{2}\right)^4 = \left(i - \sqrt{2}\right)^2\left(i - \sqrt{2}\right)^2$$

$$= \left(1 - 2i\sqrt{2}\right)\left(1 - 2i\sqrt{2}\right) = 1^2 + 1\left(-2i\sqrt{2}\right) - 2i\sqrt{2}(1) - 2i\sqrt{2}\left(-2i\sqrt{2}\right)$$

Answer Key

$= 1 - 2i\sqrt{2} - 2i\sqrt{2} + 4(2)i^2 = 1 - 4i\sqrt{2} - 8 = -7 - 4i\sqrt{2}$ such that

$(i - \sqrt{2})^8 = (i - \sqrt{2})^4 (i - \sqrt{2})^4 = (-7 - 4i\sqrt{2})(-7 - 4i\sqrt{2})$

$= (7 + 4i\sqrt{2})(7 + 4i\sqrt{2}) = 7(7) + 7(4i\sqrt{2}) + 4i\sqrt{2}(7) + 4i\sqrt{2}(4i\sqrt{2})$

$= 49 + 28i\sqrt{2} + 28i\sqrt{2} + 16(2)i^2 = 49 + 56i\sqrt{2} - 32 = 17 + 56i\sqrt{2}$

(6) Let $z = x + iy = \sqrt{3} - 1 - \frac{i}{2}$, such that $x = \sqrt{3} - 1$ and $y = -\frac{1}{2}$ (Quad. IV).

Note: It's in Quadrant IV because $\sqrt{3} - 1 > 0$ (since $\sqrt{3} > 1$).

$|z| = \sqrt{x^2 + y^2} = \sqrt{(\sqrt{3} - 1)^2 + \left(-\frac{1}{2}\right)^2} \approx \sqrt{0.535898 + 0.25} \approx 0.8865091$

$\theta = \tan^{-1}\left(\frac{y}{x}\right) = \tan^{-1}\left(\frac{-1/2}{\sqrt{3} - 1}\right) \approx \tan^{-1}\left(\frac{-0.5}{0.7320508}\right) \approx -34.3336°$ (Quad. IV)

$(|z|\cos\theta + i|z|\sin\theta)^3 = |z|^3 \cos(3\theta) + i|z|^3 \sin(3\theta)$

$\left(\sqrt{3} - 1 - \frac{i}{2}\right)^3 \approx (0.8865091)^3 \cos[3(-34.3336°)] + (0.8865091)^3 i \sin[3(-34.3336°)]$

$\left(\sqrt{3} - 1 - \frac{i}{2}\right)^3 \approx 0.696706 \cos(-103.0008°) + 0.696706 i \sin(-103.0008°)$

$\left(\sqrt{3} - 1 - \frac{i}{2}\right)^3 \approx -0.15673 - 0.67885i$

Check: $\left(\sqrt{3} - 1 - \frac{i}{2}\right)^2 \approx (0.7320508 - 0.5i)(0.7320508 - 0.5i)$

$\approx 0.5358984 - 0.3660254i - 0.3660254i + 0.25i^2$

$\approx 0.5358984 - 0.7320508i - 0.25 \approx 0.2858984 - 0.7320508i$

$\left(\sqrt{3} - 1 - \frac{i}{2}\right)^3 = \left(\sqrt{3} - 1 - \frac{i}{2}\right)^2 \left(\sqrt{3} - 1 - \frac{i}{2}\right)$

$\approx (0.2858984 - 0.7320508i)(0.7320508 - 0.5i)$

$\approx 0.209292 - 0.1429492i - 0.5358984i + 0.3660254i^2 \approx -0.15673 - 0.67885i$

If you multiply this out without rounding to decimals, if you don't make any mistakes in the arithmetic, the **exact answer** is $-\frac{37}{4} + \frac{21}{4}\sqrt{3} - \frac{47}{8}i + 3i\sqrt{3}$.

(7) Let $z = x + iy = 5 + 12i$, such that $x = 5$ and $y = 12$ (Quad. I).

$|z| = \sqrt{x^2 + y^2} = \sqrt{5^2 + 12^2} = \sqrt{25 + 144} = \sqrt{169} = 13$

$\theta = \tan^{-1}\left(\frac{y}{x}\right) = \tan^{-1}\left(\frac{12}{5}\right) \approx 67.38014°$ (Quad. I)

$$\pm\sqrt{x+iy} = \sqrt{|z|}\cos\frac{\theta}{2} + i\sqrt{|z|}\sin\frac{\theta}{2}$$

$$\pm\sqrt{5+12i} \approx \sqrt{13}\cos\left(\frac{67.38014°}{2}\right) + i\sqrt{13}\sin\left(\frac{67.38014°}{2}\right)$$

Note: Many students who get the **wrong answer** accidentally use 13 instead of $\sqrt{13}$.

$$\pm\sqrt{5+12i} \approx \sqrt{13}\cos 33.69007° + i\sqrt{13}\sin 33.69007° \approx 3 + 2i$$

$$\sqrt{5+12i} = -3 - 2i \quad \text{or} \quad \sqrt{5+12i} = 3 + 2i$$

Check: $(3+2i)(3+2i) = 3(3) + 3(2i) + 2i(3) + 2i(2i) = 9 + 6i + 6i + 4i^2$
$= 9 + 12i - 4 = 5 + 12i$

(8) Let $z = x + iy = 0 + i$, such that $x = 0$ and $y = 1$ (on the +y-axis).

$$|z| = \sqrt{x^2 + y^2} = \sqrt{0^2 + 1^2} = \sqrt{0+1} = \sqrt{1} = 1$$

$$\theta = \tan^{-1}\left(\frac{y}{x}\right) = \tan^{-1}\left(\frac{1}{0}\right) = 90° \text{ (on the +y-axis)}$$

Note: See the solution to Exercise 4 in Chapter 7. Although $\frac{1}{0}$ is undefined, we know that $\tan^{-1}\left(\frac{1}{0}\right) = 90°$ since the point $(0, i)$ lies directly on the +y-axis.

$$\pm\sqrt{x+iy} = \sqrt{|z|}\cos\frac{\theta}{2} + i\sqrt{|z|}\sin\frac{\theta}{2}$$

$$\pm\sqrt{i} \approx \sqrt{1}\cos\left(\frac{90°}{2}\right) + i\sqrt{1}\sin\left(\frac{90°}{2}\right) = \cos 45° + i\sin 45° = \frac{\sqrt{2}}{2} + \frac{\sqrt{2}}{2}i$$

$$\sqrt{i} = -\frac{\sqrt{2}}{2} - \frac{\sqrt{2}}{2}i \quad \text{or} \quad \sqrt{i} = \frac{\sqrt{2}}{2} + \frac{\sqrt{2}}{2}i$$

Alternate answer: $\sqrt{i} = -\frac{1}{\sqrt{2}} - \frac{1}{\sqrt{2}}i$ or $\sqrt{i} = \frac{1}{\sqrt{2}} + \frac{1}{\sqrt{2}}i$ (since $\frac{1}{\sqrt{2}} = \frac{1}{\sqrt{2}}\frac{\sqrt{2}}{\sqrt{2}} = \frac{\sqrt{2}}{2}$)

Check: $\left(\frac{\sqrt{2}}{2} + \frac{\sqrt{2}}{2}i\right)\left(\frac{\sqrt{2}}{2} + \frac{\sqrt{2}}{2}i\right) = \frac{\sqrt{2}\sqrt{2}}{2\cdot 2} + \frac{\sqrt{2}\sqrt{2}}{2\cdot 2}i + \frac{\sqrt{2}}{2}i\frac{\sqrt{2}}{2} + \frac{\sqrt{2}\sqrt{2}}{2\cdot 2}i^2$

$= \frac{2}{4} + \frac{2}{4}i + \frac{2}{4}i + \frac{2}{4}(-1) = \left(\frac{2}{4} - \frac{2}{4}\right) + \left(\frac{2}{4} + \frac{2}{4}\right)i = 0 + \frac{4}{4}i = i$

(9) Let $z = x + iy = 0 - i$, such that $x = 0$ and $y = -1$ (on the $-y$-axis).

$$|z| = \sqrt{x^2 + y^2} = \sqrt{0^2 + (-1)^2} = \sqrt{0+1} = \sqrt{1} = 1$$

$$\theta = \tan^{-1}\left(\frac{y}{x}\right) = \tan^{-1}\left(\frac{-1}{0}\right) = -90° \text{ or } 270° \text{ (on the neg. y-axis)}$$

Note: See the solution to Exercise 9 in Chapter 7. Although $\frac{-1}{0}$ is undefined, we know that $\tan^{-1}\left(\frac{-1}{0}\right) = 270°$ since the point $(0, -1)$ lies directly on the $-y$-axis.

Answer Key

$$\pm\sqrt{x+iy} = \sqrt{|z|}\cos\frac{\theta}{2} + i\sqrt{|z|}\sin\frac{\theta}{2}$$

$$\pm\sqrt{-i} \approx \sqrt{1}\cos\left(\frac{270°}{2}\right) + i\sqrt{1}\sin\left(\frac{270°}{2}\right) = \cos 135° + i\sin 135° = -\frac{\sqrt{2}}{2} + \frac{\sqrt{2}}{2}i$$

$$\sqrt{-i} = \frac{\sqrt{2}}{2} - \frac{\sqrt{2}}{2}i \quad \text{or} \quad \sqrt{-i} = -\frac{\sqrt{2}}{2} + \frac{\sqrt{2}}{2}i$$

Alternate answer: $\sqrt{-i} = \frac{1}{\sqrt{2}} - \frac{1}{\sqrt{2}}i$ or $\sqrt{-i} = -\frac{1}{\sqrt{2}} + \frac{1}{\sqrt{2}}i$ (since $\frac{1}{\sqrt{2}} = \frac{1}{\sqrt{2}}\frac{\sqrt{2}}{\sqrt{2}} = \frac{\sqrt{2}}{2}$)

Check: $\left(\frac{\sqrt{2}}{2} - \frac{\sqrt{2}}{2}i\right)\left(\frac{\sqrt{2}}{2} - \frac{\sqrt{2}}{2}i\right) = \frac{\sqrt{2}}{2}\frac{\sqrt{2}}{2} - \frac{\sqrt{2}}{2}\frac{\sqrt{2}}{2}i - \frac{\sqrt{2}}{2}i\frac{\sqrt{2}}{2} + \frac{\sqrt{2}}{2}\frac{\sqrt{2}}{2}i^2$

$= \frac{2}{4} - \frac{2}{4}i - \frac{2}{4}i + \frac{2}{4}(-1) = \left(\frac{2}{4} - \frac{2}{4}\right) + \left(-\frac{2}{4} - \frac{2}{4}\right)i = 0 - \frac{4}{4}i = -i$

(10) Let $z = x + iy = 15 - 8i$, such that $x = 15$ and $y = -8$ (Quad. IV).

$$|z| = \sqrt{x^2 + y^2} = \sqrt{15^2 + (-8)^2} = \sqrt{225 + 64} = \sqrt{289} = 17$$

$$\theta = \tan^{-1}\left(\frac{y}{x}\right) = \tan^{-1}\left(\frac{-8}{15}\right) \approx -28.07249° \text{ or } 331.92752° \text{ (Quad. IV)}$$

$$\pm\sqrt{x+iy} = \sqrt{|z|}\cos\frac{\theta}{2} + i\sqrt{|z|}\sin\frac{\theta}{2}$$

$$\pm\sqrt{15 - 8i} \approx \sqrt{17}\cos\left(\frac{-28.07249°}{2}\right) + i\sqrt{17}\sin\left(\frac{-28.07249°}{2}\right)$$

Note: Many students who get the **wrong answer** accidentally use 17 instead of $\sqrt{17}$.

$$\pm\sqrt{15 - 8i} \approx \sqrt{17}\cos(-14.03625°) + i\sqrt{17}\sin(-14.03625°) \approx 4 - i$$

$$\sqrt{15 - 8i} = -4 + i \quad \text{or} \quad \sqrt{15 - 8i} = 4 - i$$

Check: $(4 - i)(4 - i) = 4(4) + 4(-i) - i(4) + i^2 = 16 - 4i - 4i - 1$
$= 15 - 8i$

(11) Let $z = x + iy = 28i - 45 = -45 + 28i$, such that $x = -45$ and $y = 28$ (Quad. II).

$$|z| = \sqrt{x^2 + y^2} = \sqrt{(-45)^2 + 28^2} = \sqrt{2025 + 784} = \sqrt{2809} = 53$$

$$\theta = \tan^{-1}\left(\frac{y}{x}\right) = \tan^{-1}\left(\frac{28}{-45}\right) \approx -31.89079° + 180° \approx 148.10921° \text{ (Quad. II)}$$

Note: Add 180° to the calculator's answer for inverse tangent if $x < 0$ (Chapter 7).

$$\pm\sqrt{x+iy} = \sqrt{|z|}\cos\frac{\theta}{2} + i\sqrt{|z|}\sin\frac{\theta}{2}$$

$$\pm\sqrt{28i - 45} \approx \sqrt{53}\cos\left(\frac{148.10921°}{2}\right) + i\sqrt{53}\sin\left(\frac{148.10921°}{2}\right)$$

Note: Many students who get the **wrong answer** accidentally use 53 instead of $\sqrt{53}$.

Complex Numbers Essentials Math Workbook with Answers

$$\pm\sqrt{28i-45} \approx \sqrt{53}\cos 74.05461° + i\sqrt{53}\sin 74.05461° \approx 2 + 7i$$
$$\sqrt{28i-45} = \pm(2+7i)$$
$$\sqrt{28i-45} = 2+7i \quad \text{or} \quad \sqrt{28i-45} = -2-7i$$

Check: $(2+7i)(2+7i) = 2(2) + 2(7i) + 7i(2) + 7i(7i) = 4 + 14i + 14i + 49i^2 = 4 + 28i - 49 = -45 + 28i = 28i - 45$

(12) Let $z = x + iy = 1 + i\sqrt{3}$, such that $x = 1$ and $y = \sqrt{3}$ (Quad. I).

$$|z| = \sqrt{x^2 + y^2} = \sqrt{1^2 + (\sqrt{3})^2} = \sqrt{1+3} = \sqrt{4} = 2$$

$$\theta = \tan^{-1}\left(\frac{y}{x}\right) = \tan^{-1}\left(\frac{\sqrt{3}}{1}\right) = \tan^{-1}(\sqrt{3}) = 60° \text{ (Quad. I)}$$

$$\pm\sqrt{x+iy} = \sqrt{|z|}\cos\frac{\theta}{2} + i\sqrt{|z|}\sin\frac{\theta}{2}$$

$$\pm\sqrt{1+i\sqrt{3}} = \sqrt{2}\cos\left(\frac{60°}{2}\right) + i\sqrt{2}\sin\left(\frac{60°}{2}\right)$$

Note: Many students who get the **wrong answer** accidentally use 2 instead of $\sqrt{2}$.

$$\pm\sqrt{1+i\sqrt{3}} = \sqrt{2}\cos 30° + i\sqrt{2}\sin 30° = \sqrt{2}\left(\frac{\sqrt{3}}{2}\right) + i\sqrt{2}\left(\frac{1}{2}\right)$$

$$\pm\sqrt{1+i\sqrt{3}} = \frac{\sqrt{(2)(3)}}{2} + \frac{\sqrt{2}}{2}i = \frac{\sqrt{6}}{2} + \frac{\sqrt{2}}{2}i$$

$$\sqrt{1+i\sqrt{3}} = \frac{\sqrt{6}}{2} + \frac{\sqrt{2}}{2}i \quad \text{or} \quad \sqrt{1+i\sqrt{3}} = -\frac{\sqrt{6}}{2} - \frac{\sqrt{2}}{2}i$$

Alternate answers: $\pm\sqrt{1+i\sqrt{3}} = \frac{\sqrt{3}}{\sqrt{2}} + \frac{\sqrt{2}}{2}i = \frac{\sqrt{3}}{\sqrt{2}} + \frac{1}{\sqrt{2}}i = \sqrt{\frac{3}{2}} + \frac{\sqrt{2}}{2}i = \sqrt{\frac{3}{2}} + \frac{1}{\sqrt{2}}i$

Check: $\left(\frac{\sqrt{6}}{2} + \frac{\sqrt{2}}{2}i\right)\left(\frac{\sqrt{6}}{2} + \frac{\sqrt{2}}{2}i\right) = \frac{\sqrt{6}\sqrt{6}}{2\cdot 2} + \frac{\sqrt{6}\sqrt{2}}{2\cdot 2}i + \frac{\sqrt{2}}{2}i\frac{\sqrt{6}}{2} + \frac{\sqrt{2}\sqrt{2}}{2\cdot 2}i^2 = \frac{6}{4} + \frac{\sqrt{12}}{4}i + \frac{\sqrt{12}}{4}i + \frac{2}{4}(-1)$

$= \frac{6}{4} - \frac{2}{4} + \frac{2\sqrt{12}}{4}i = \frac{4}{4} + \frac{\sqrt{12}}{2}i = 1 + \frac{\sqrt{12}}{2} = 1 + \frac{\sqrt{(4)(3)}}{2} = 1 + \frac{\sqrt{4}\sqrt{3}}{2} = 1 + \frac{2}{2}\sqrt{3} = 1 + \sqrt{3}$

Answer Key

Chapter 11

(1) $e^{i\pi/3} = \cos\left(\frac{\pi}{3}\right) + i\sin\left(\frac{\pi}{3}\right) = \frac{1}{2} + \frac{\sqrt{3}}{2}i$ Note: $\frac{\pi}{3}$ rad $= \frac{\pi}{3}\frac{180°}{\pi} = \frac{180°}{3} = 60°$

(2) $e^{i\pi/4} = \cos\left(\frac{\pi}{4}\right) + i\sin\left(\frac{\pi}{4}\right) = \frac{\sqrt{2}}{2} + \frac{\sqrt{2}}{2}i$ Note: $\frac{\pi}{4}$ rad $= \frac{\pi}{4}\frac{180°}{\pi} = \frac{180°}{4} = 45°$

Alternate answer: $\frac{1}{\sqrt{2}} + \frac{1}{\sqrt{2}}i$ since $\frac{1}{\sqrt{2}} = \frac{1}{\sqrt{2}}\frac{\sqrt{2}}{\sqrt{2}} = \frac{\sqrt{2}}{2}$

(3) $e^{3i\pi/2} = \cos\left(\frac{3\pi}{2}\right) + i\sin\left(\frac{3\pi}{2}\right) = 0 - 1i = -i$ Note: $\frac{3\pi}{2}$ rad $= \frac{3\pi}{2}\frac{180°}{\pi} = \frac{3(180°)}{2} = 270°$

(4) $e^{2i\pi/3} = \cos\left(\frac{2\pi}{3}\right) + i\sin\left(\frac{2\pi}{3}\right) = -\frac{1}{2} + \frac{\sqrt{3}}{2}i$ Note: $\frac{2\pi}{3}$ rad $= \frac{2\pi}{3}\frac{180°}{\pi} = \frac{2(180°)}{3} = 120°$

(5) $e^{7i\pi/6} = \cos\left(\frac{7\pi}{6}\right) + i\sin\left(\frac{7\pi}{6}\right) = -\frac{\sqrt{3}}{2} - \frac{i}{2}$ Note: $\frac{7\pi}{6}$ rad $= \frac{7\pi}{6}\frac{180°}{\pi} = \frac{7(180°)}{6} = 210°$

(6) $e^{-i\pi/3} = \cos\left(-\frac{\pi}{3}\right) + i\sin\left(-\frac{\pi}{3}\right) = \frac{1}{2} - \frac{\sqrt{3}}{2}i$ Note: $-\frac{\pi}{3}$ rad $= -\frac{\pi}{3}\frac{180°}{\pi} = -\frac{180°}{3} = -60°$

(7) $e^{i\pi/12}e^{i\pi/4} = e^{i\pi/3} = \frac{1}{2} + \frac{\sqrt{3}}{2}i$ Notes: We used the rule $e^a e^b = e^{a+b}$.

$\frac{1}{12} + \frac{1}{4} = \frac{1}{12} + \frac{1}{4}\frac{3}{3} = \frac{1}{12} + \frac{3}{12} = \frac{4}{12} = \frac{1}{3}$. Recall from Exercise 1 that $e^{i\pi/3} = \frac{1}{2} + \frac{\sqrt{3}}{2}i$.

(8) $\frac{e^{13i\pi/30}}{e^{3i\pi/5}} = e^{-i\pi/6} = \cos\left(-\frac{\pi}{6}\right) + i\sin\left(-\frac{\pi}{6}\right) = \frac{\sqrt{3}}{2} - \frac{i}{2}$ Notes: We used the rule $\frac{e^a}{e^b} = e^{a-b}$.

$\frac{13}{30} - \frac{3}{5} = \frac{13}{30} - \frac{3}{5}\frac{6}{6} = \frac{13}{30} - \frac{18}{30} = -\frac{5}{30} = -\frac{1}{6}$. Also, $-\frac{\pi}{6}$ rad $= -\frac{\pi}{6}\frac{180°}{\pi} = -\frac{180°}{6} = -30°$.

(9) $\left(e^{i\pi/15}\right)^5 = e^{i\pi/3} = \frac{1}{2} + \frac{\sqrt{3}}{2}i$ Notes: We used the rule $(e^a)^b = e^{ab}$.

$\left(\frac{1}{15}\right)(5) = \frac{5}{15} = \frac{1}{3}$. Recall from Exercise 1 that $e^{i\pi/3} = \frac{1}{2} + \frac{\sqrt{3}}{2}i$.

(10) $\left(\cos\frac{5\pi}{4} + i\sin\frac{5\pi}{4}\right)\sqrt{2} - i\left(\cos\frac{\pi}{3} + i\sin\frac{\pi}{3}\right) = \left(-\frac{\sqrt{2}}{2} - \frac{\sqrt{2}}{2}i\right)\sqrt{2} - i\left(\frac{1}{2} + \frac{\sqrt{3}}{2}i\right)$

$-\frac{\sqrt{2}\sqrt{2}}{2} - \frac{\sqrt{2}\sqrt{2}}{2}i - \frac{i}{2} - i^2\frac{\sqrt{3}}{2} = -\frac{2}{2} - \frac{2}{2}i - \frac{i}{2} - (-1)\frac{\sqrt{3}}{2} = -1 - \frac{3i}{2} + \frac{\sqrt{3}}{2} = \left(\frac{\sqrt{3}}{2} - 1\right) - \frac{3i}{2}$

Notes: $\frac{\pi}{3}$ rad $= \frac{\pi}{3}\frac{180°}{\pi} = \frac{180°}{3} = 60°$ and $\frac{5\pi}{4}$ rad $= \frac{5\pi}{4}\frac{180°}{\pi} = \frac{5(180°)}{4} = \frac{900°}{4} = 225°$.

(11) First, recall from Exercise 2 that $e^{i\pi/4} = \frac{\sqrt{2}}{2} + \frac{\sqrt{2}}{2}i$. After we make this substitution, we will factor $\sqrt{2}$ out of both the numerator and denominator, and then we will multiply each by 2.

$\frac{\sqrt{2}+e^{i\pi/4}}{\sqrt{2}-e^{i\pi/4}} = \frac{\sqrt{2}+\frac{\sqrt{2}}{2}+\frac{\sqrt{2}}{2}i}{\sqrt{2}-\frac{\sqrt{2}}{2}+\frac{\sqrt{2}}{2}i} = \frac{\sqrt{2}\left(1+\frac{1}{2}+\frac{i}{2}\right)}{\sqrt{2}\left(1-\frac{1}{2}+\frac{i}{2}\right)} = \frac{1+\frac{1}{2}+\frac{i}{2}}{1-\frac{1}{2}+\frac{i}{2}} = \frac{\left(1+\frac{1}{2}+\frac{i}{2}\right)2}{\left(1-\frac{1}{2}+\frac{i}{2}\right)2} = \frac{2+1+i}{2-1+i} = \frac{3+i}{1+i}$

Now divide the fractions like we did in Chapter 4.

Complex Numbers Essentials Math Workbook with Answers

$$\frac{3+i}{1+i}\left(\frac{1-i}{1-i}\right) = \frac{3-3i+i-i^2}{1+1} = \frac{3-2i-(-1)}{2} = \frac{3-2i+1}{2} = \frac{4-2i}{2} = \frac{4}{2} - \frac{2i}{2} = 2-i$$

(12) Recall from Exercise 1 that $e^{i\pi/3} = \frac{1}{2} + \frac{\sqrt{3}}{2}i$. We will combine the imaginary terms (which are like terms) together by finding a common denominator.

$$\left(i\sqrt{3} + e^{i\pi/3}\right)^2 = \left(i\sqrt{3} + \frac{1}{2} + \frac{\sqrt{3}}{2}i\right)^2 = \left(\frac{2i\sqrt{3}}{2} + \frac{1}{2} + \frac{\sqrt{3}}{2}i\right)^2 = \left(\frac{3i\sqrt{3}}{2} + \frac{1}{2}\right)^2$$

$$= \left(\frac{3i\sqrt{3}}{2} + \frac{1}{2}\right)\left(\frac{3i\sqrt{3}}{2} + \frac{1}{2}\right) = \frac{3i\sqrt{3}}{2}\left(\frac{3i\sqrt{3}}{2}\right) + \frac{3i\sqrt{3}}{2}\left(\frac{1}{2}\right) + \frac{1}{2}\left(\frac{3i\sqrt{3}}{2}\right) + \frac{1}{2}\left(\frac{1}{2}\right)$$

$$= \frac{9i^2(3)}{4} + \frac{3i\sqrt{3}}{4} + \frac{3i\sqrt{3}}{4} + \frac{1}{4} = -\frac{27}{4} + \frac{6i\sqrt{3}}{4} + \frac{1}{4} = -\frac{26}{4} + \frac{3i\sqrt{3}}{2} = -\frac{13}{2} + \frac{3i\sqrt{3}}{2}$$

(13) Compare $7 + 7i$ with $x + iy$ to see that $x = 7$ and $y = 7$ (Quad. I).

$$|z| = \sqrt{x^2 + y^2} = \sqrt{7^2 + 7^2} = \sqrt{49 + 49} = \sqrt{98} = \sqrt{49(2)} = \sqrt{49}\sqrt{2} = 7\sqrt{2}$$

$$\theta = \tan^{-1}\left(\frac{y}{x}\right) = \tan^{-1}\left(\frac{7}{7}\right) = \tan^{-1}(1) = 45° = 45\frac{\pi}{180} = \frac{\pi}{4} \text{ (Quad. I)}$$

The exponential form is $7\sqrt{2}e^{i\pi/4}$ or $7\sqrt{2}e^{i\pi/4 + 2i\pi k}$.

Note: **Since the trig functions are periodic, we may add $2i\pi k$ to the exponent where k is an integer and obtain an equivalent answer.**

Note: Convert $45°$ to $\frac{\pi}{4}$ **radians** before expressing the answer in exponential form.

(14) Compare $\sqrt{2} - i\sqrt{6}$ with $x + iy$ to see that $x = \sqrt{2}$ and $y = -\sqrt{6}$ (Quad. IV).

$$|z| = \sqrt{x^2 + y^2} = \sqrt{\left(\sqrt{2}\right)^2 + \left(\sqrt{6}\right)^2} = \sqrt{2+6} = \sqrt{8} = \sqrt{4(2)} = \sqrt{4}\sqrt{2} = 2\sqrt{2}$$

$$\theta = \tan^{-1}\left(\frac{y}{x}\right) = \tan^{-1}\left(\frac{-\sqrt{6}}{\sqrt{2}}\right) = \tan^{-1}\left(-\sqrt{\frac{6}{2}}\right) = \tan^{-1}(-\sqrt{3})$$

$$\theta = -60° = -60\frac{\pi}{180} = -\frac{\pi}{3} \text{ (Quad. IV)}$$

The exponential form is $2\sqrt{2}e^{-i\pi/3}$ or $2\sqrt{2}e^{-i\pi/3 + 2i\pi k}$.

Alternate answer: $-60°$ is equivalent to $-60° + 360° = 300°$, which equates to $300° = 300\frac{\pi}{180} = \frac{5\pi}{3}$. This gives the alternate answer $2\sqrt{2}e^{5i\pi/3}$.

(15) Compare $27i$ with $x + iy$ to see that $x = 0$ and $y = 27$ (on the $+y$-axis).

$$|z| = \sqrt{x^2 + y^2} = \sqrt{0^2 + 27^2} = \sqrt{27^2} = 27$$

$$\theta = \tan^{-1}\left(\frac{y}{x}\right) = \tan^{-1}\left(\frac{27}{0}\right) = 90° = 90\frac{\pi}{180} = \frac{\pi}{2} \text{ (on the } +y\text{-axis)}$$

Answer Key

The exponential form is $27e^{i\pi/2}$ or $27e^{i\pi/2+2i\pi k}$. Note: See the solution to Exercise 4 in Chapter 7. Although $\frac{27}{0}$ is undefined, we know that $\tan^{-1}\left(\frac{27}{0}\right) = 90°$ since the point $(0, 27i)$ lies directly on the $+y$-axis.

(16) Compare $i - \sqrt{3} = -\sqrt{3} + i$ with $x + iy$ to see that $x = -\sqrt{3}$ and $y = 1$ (Quad. II).

$$|z| = \sqrt{x^2 + y^2} = \sqrt{\left(-\sqrt{3}\right)^2 + 1^2} = \sqrt{3+1} = \sqrt{4} = 2$$

$$\theta = \tan^{-1}\left(\frac{y}{x}\right) = \tan^{-1}\left(\frac{1}{-\sqrt{3}}\right) = -30° + 180° = 150° \text{ (Quad. II)}$$

Note: Add 180° to the calculator's answer for inverse tangent if $x < 0$ (Chapter 7).

The exponential form is $2e^{5i\pi/6}$ or $2e^{5i\pi/6+2i\pi k}$. Note: $150° = 150\frac{\pi}{180} = \frac{5\pi}{6}$.

(17) Compare $\frac{5}{i} = \frac{5\,i}{i\,i} = \frac{5i}{i^2} = \frac{5i}{-1} = -5i$ with $x + iy$ to see that $x = 0$ and $y = -5$ (on the $-y$-axis).

$$|z| = \sqrt{x^2 + y^2} = \sqrt{0^2 + (-5)^2} = \sqrt{25} = 5$$

$$\theta = \tan^{-1}\left(\frac{y}{x}\right) = \tan^{-1}\left(\frac{-5}{0}\right) = 270° = 270\frac{\pi}{180} = \frac{3\pi}{2} \text{ (on the neg. }y\text{-axis)}$$

The exponential form is $5e^{3i\pi/2}$ or $5e^{3i\pi/2+2i\pi k}$. Note: See the solution to Exercise 9 in Chapter 7. Although $\frac{-5}{0}$ is undefined, we know that $\tan^{-1}\left(\frac{-5}{0}\right) = 270°$ since the point $(0, -5i)$ lies directly on the $-y$-axis.

Alternate answer: 270° is equivalent to $-90°$, which equates to $-90° = -90\frac{\pi}{180} = -\frac{\pi}{2}$. This gives the alternate answer $5e^{-i\pi/2}$.

(18) Recall from Chapter 3 that an overbar indicates a complex conjugate:
$\overline{2 + 2i\sqrt{3}} = 2 - i2\sqrt{3}$. Compare $2 - i2\sqrt{3}$ with $x + iy$ to see that $x = 2$ and $y = -2\sqrt{3}$ (Quad. IV).

$$|z| = \sqrt{x^2 + y^2} = \sqrt{2^2 + \left(-2\sqrt{3}\right)^2} = \sqrt{4 + (-2)^2\left(\sqrt{3}\right)^2} = \sqrt{4 + 12} = \sqrt{16} = 4$$

$$\theta = \tan^{-1}\left(\frac{y}{x}\right) = \tan^{-1}\left(\frac{-2\sqrt{3}}{2}\right) = \tan^{-1}\left(-\sqrt{3}\right) = -60° = -60\frac{\pi}{180} = -\frac{\pi}{3} \text{ (Quad. IV)}$$

The exponential form is $4e^{-i\pi/3}$ or $4e^{-i\pi/3+2i\pi k}$.

Alternate answer: $-60°$ is equivalent to $-60° + 360° = 300°$, which equates to $300° = 300\frac{\pi}{180} = \frac{5\pi}{3}$. This gives the alternate answer $4e^{5i\pi/3}$.

(19) First find $(2 + 3i)(2 - 3i) = 2(2) + 2(-3i) + 3i(2) + 3i(-3i)$
$= 4 - 6i + 6i - 9i^2 = 4 - 9(-1) = 4 + 9 = 13$. Alternatively, just use the formula for multiplying a number by its complex conjugate (Chapter 3). The number 13 is purely real. (When you multiply a number by its complex conjugate, you always get a real number.) Here, $x = 13$, $y = 0$, $|z| = 13$, and $\theta = 0$, such that the exponential form is $13e^0 = 13(1) = 13$. If the number is **purely real**, the Cartesian form and exponential form are the same (each equals x). Note that $13 = 13e^{2i\pi k}$.

(20) First carry out the division according to Chapter 4:

$$\frac{15i-9}{4-i} = \frac{15i-9}{4-i}\left(\frac{4+i}{4+i}\right) = \frac{60i+15i^2-36-9i}{4^2+1^2} = \frac{51i-15-36}{16+1} = \frac{51i-51}{17} = \frac{51i}{17} - \frac{51}{17} = 3i - 3$$

Compare $3i - 3 = -3 + 3i$ with $x + iy$ to see that $x = -3$ and $y = 3$ (Quad. II).

$$|z| = \sqrt{x^2 + y^2} = \sqrt{(-3)^2 + 3^2} = \sqrt{9+9} = \sqrt{18} = \sqrt{9(2)} = \sqrt{9}\sqrt{2} = 3\sqrt{2}$$

$$\theta = \tan^{-1}\left(\frac{y}{x}\right) = \tan^{-1}\left(\frac{3}{-3}\right) = \tan^{-1}(-1) = -45° + 180° = 135° \text{ (Quad. II)}$$

Note: Add 180° to the calculator's answer for inverse tangent if $x < 0$ (Chapter 7).

$$135° = 135\frac{\pi}{180} = \frac{3\pi}{4}$$

The exponential form is $3\sqrt{2}e^{3i\pi/4}$ or $3\sqrt{2}e^{3i\pi/4+2i\pi k}$.

(21) Note: One way to solve this problem is like Exercise 9 in Chapter 10. It's practically the same problem, except for the 9 and converting the answer to exponential form. We'll show another way to approach square roots here. Let $z = -9i = x + iy$, such that $x = 0$ and $y = -9$; we'll worry about the square root later.

$$|z| = \sqrt{x^2 + y^2} = \sqrt{0^2 + (-9)^2} = \sqrt{81} = 9$$

$$\theta = \tan^{-1}\left(\frac{y}{x}\right) = \tan^{-1}\left(\frac{-9}{0}\right) = 270° = 270\frac{\pi}{180} = \frac{3\pi}{2} \text{ (on the neg. } y\text{-axis)}$$

(Alternatively, you could use $\theta = -90° = -\frac{\pi}{2}$.) This gives us $z = 9e^{3i\pi/2}$, but we're not finished yet because we defined $z = -9i$, but we really want $\sqrt{-9i}$. We need to deal with the square root. Recall from algebra that $t^{1/2} = \sqrt{t}$. This allows us to write $\sqrt{-9i} = (-9i)^{1/2}$. Since we let $z = -9i$ and we found that $z = 9e^{3i\pi/2}$, what we need is $(9e^{3i\pi/2})^{1/2}$. Apply the rule $(ab)^c = a^c b^c$ to get $9^{1/2}(e^{3i\pi/2})^{1/2}$. Apply the rule $(e^u)^v = e^{uv}$ to get $9^{1/2}e^{3i\pi/4} = 3e^{3i\pi/4}$. Since square roots have two answers (for

Answer Key

example, i and $-i$ both satisfy $\sqrt{-1}$), there is actually a second answer. As we'll explore in Chapter 13, the second answer is $3e^{7i\pi/4}$ (equivalent to $3e^{-i\pi/4}$), since $\frac{3i\pi}{4}$ rad and $\frac{7i\pi}{4}$ rad are separated by π rad (or 180°). As usual, you can express the general answer in terms of k, allowing for the periodicity of the trig functions. But should it be $2i\pi k$ or something else? Consider the square root carefully.

Check the answer: $u = 3e^{3i\pi/4}$ has $|u| = 3$ and $\varphi = \frac{3\pi}{4}$. The Cartesian components are $|u|\cos\frac{3\pi}{4} = (3)\left(-\frac{\sqrt{2}}{2}\right) = -\frac{3\sqrt{2}}{2}$ and $|u|\sin\frac{3\pi}{4} = (3)\left(\frac{\sqrt{2}}{2}\right) = \frac{3\sqrt{2}}{2}$. This means that $\sqrt{-9i} = -\frac{3\sqrt{2}}{2} + \frac{3\sqrt{2}}{2}i$. To check this answer, multiply $-\frac{3\sqrt{2}}{2} + \frac{3\sqrt{2}}{2}i$ by itself and see if that equals $-9i$. We get $\left(-\frac{3\sqrt{2}}{2} + \frac{3\sqrt{2}}{2}i\right)\left(-\frac{3\sqrt{2}}{2} + \frac{3\sqrt{2}}{2}i\right) = -\frac{3\sqrt{2}}{2}\left(-\frac{3\sqrt{2}}{2}\right) - \frac{3\sqrt{2}}{2}\left(\frac{3\sqrt{2}}{2}i\right) + \frac{3\sqrt{2}}{2}i\left(-\frac{3\sqrt{2}}{2}\right) + \frac{3\sqrt{2}}{2}i\left(\frac{3\sqrt{2}}{2}i\right) = \frac{9(2)}{4} - \frac{9(2)}{4}i - \frac{9(2)}{4}i + \frac{9(2)}{4}i^2 = \frac{9}{2} - \frac{9}{2}i - \frac{9}{2}i + \frac{9}{2}(-1) = -9i$.

(22) Compare $8e^{i\pi/6}$ with $|z|e^{i\theta}$ to see that $|z| = 8$ and $\theta = \frac{\pi}{6}$.

$x = |z|\cos\theta = 8\cos\frac{\pi}{6} = 8\left(\frac{\sqrt{3}}{2}\right) = 4\sqrt{3}$ and $y = |z|\sin\theta = 8\sin\frac{\pi}{6} = 8\left(\frac{1}{2}\right) = 4$

The Cartesian form is $4\sqrt{3} + 4i$. Note: $\frac{\pi}{6} = \frac{\pi}{6}\frac{180°}{\pi} = \frac{180°}{6} = 30°$.

(23) Compare $10e^{2i\pi/3}$ with $|z|e^{i\theta}$ to see that $|z| = 10$ and $\theta = \frac{2\pi}{3}$.

$x = |z|\cos\theta = 10\cos\frac{2\pi}{3} = 10\left(-\frac{1}{2}\right) = -5$

$y = |z|\sin\theta = 10\sin\frac{2\pi}{3} = 10\left(\frac{\sqrt{3}}{2}\right) = 5\sqrt{3}$

The Cartesian form is $-5 + 5i\sqrt{3}$. Note: $\frac{2\pi}{3} = \frac{2\pi}{3}\frac{180°}{\pi} = \frac{2(180°)}{3} = 120°$.

(24) Compare $\sqrt{3}e^{11i\pi/6}$ with $|z|e^{i\theta}$ to see that $|z| = \sqrt{3}$ and $\theta = \frac{11\pi}{6}$.

$x = |z|\cos\theta = \sqrt{3}\cos\frac{11\pi}{6} = \sqrt{3}\left(\frac{\sqrt{3}}{2}\right) = \frac{3}{2}$ and $y = |z|\sin\theta = \sqrt{3}\sin\frac{11\pi}{6} = \sqrt{3}\left(-\frac{1}{2}\right) = -\frac{\sqrt{3}}{2}$

The Cartesian form is $\frac{3}{2} - \frac{\sqrt{3}}{2}i$. Note: $\frac{11\pi}{6} = \frac{11\pi}{6}\frac{180°}{\pi} = \frac{11(180°)}{6} = 330°$.

(25) Compare $12e^{5i\pi/4}$ with $|z|e^{i\theta}$ to see that $|z| = 12$ and $\theta = \frac{5\pi}{4}$.

$x = |z|\cos\theta = 12\cos\frac{5\pi}{4} = 12\left(-\frac{\sqrt{2}}{2}\right) = -6\sqrt{2}$

$y = |z|\sin\theta = 12\sin\frac{5\pi}{4} = 12\left(-\frac{\sqrt{2}}{2}\right) = -6\sqrt{2}$

The Cartesian form is $-6\sqrt{2} - 6i\sqrt{2}$. Note: $\frac{5\pi}{4} = \frac{5\pi}{4}\frac{180°}{\pi} = \frac{5(180°)}{4} = \frac{900°}{4} = 225°$.

Alternate answer: $-\frac{12}{\sqrt{2}} - \frac{12i}{\sqrt{2}}$ since $\frac{12}{\sqrt{2}} = \frac{12\sqrt{2}}{\sqrt{2}\sqrt{2}} = \frac{12\sqrt{2}}{2} = 6\sqrt{2}$.

(26) Compare $9e^{i\pi/2}$ with $|z|e^{i\theta}$ to see that $|z| = 9$ and $\theta = \frac{\pi}{2}$.

$x = |z|\cos\theta = 9\cos\frac{\pi}{2} = 9(0) = 0$ and $y = |z|\sin\theta = 9\sin\frac{\pi}{2} = 9(1) = 9$

The Cartesian form is $0 + 9i = 9i$. Note: $\frac{\pi}{2} = \frac{\pi}{2}\frac{180°}{\pi} = \frac{180°}{2} = 90°$.

(27) Compare $7e^{8i\pi}$ with $|z|e^{i\theta}$ to see that $|z| = 7$ and $\theta = 8\pi$.

$x = |z|\cos\theta = 7\cos(8\pi) = 7(1) = 7$ and $y = |z|\sin\theta = 7\sin(8\pi) = 7(0) = 0$

The Cartesian form is $7 + 0i = 7$. Note: $8\pi = 8\pi\frac{180°}{\pi} = 8(180°) = 1440°$. (Since 1440 is a multiple of 360, 1440° is equivalent to 360°.)

(28) Compare $\sqrt{6}e^{5i\pi/3}$ with $|z|e^{i\theta}$ to see that $|z| = \sqrt{6}$ and $\theta = \frac{5\pi}{3}$.

$x = |z|\cos\theta = \sqrt{6}\cos\frac{5\pi}{3} = \sqrt{6}\left(\frac{1}{2}\right) = \frac{\sqrt{6}}{2}$

$y = |z|\sin\theta = \sqrt{6}\sin\frac{5\pi}{3} = \sqrt{6}\left(-\frac{\sqrt{3}}{2}\right) = -\frac{\sqrt{18}}{2} = -\frac{\sqrt{(9)(2)}}{2} = -\frac{\sqrt{9}\sqrt{2}}{2} = -\frac{3\sqrt{2}}{2}$

The Cartesian form is $\frac{\sqrt{6}}{2} - \frac{3\sqrt{2}}{2}i$. Note: $\frac{5\pi}{3} = \frac{5\pi}{3}\frac{180°}{\pi} = \frac{5(180°)}{3} = 300°$.

(29) Compare $3e^{3i\pi}$ with $|z|e^{i\theta}$ to see that $|z| = 3$ and $\theta = 3\pi$.

$x = |z|\cos\theta = 3\cos(3\pi) = 3(-1) = -3$ and $y = |z|\sin\theta = 3\sin(3\pi) = 3(0) = 0$

The Cartesian form is $-3 + 0i = -3$. Note: $3\pi = 3\pi\frac{180°}{\pi} = 3(180°) = 540°$. (Since $540 = 180 + 360$, 540° is equivalent to 180°.)

(30) Since technically a modulus is always real (Chapter 6), we should think of this as $i(\sqrt{2}e^{-i\pi/4})$ and multiply by the overall i later. In that case, we will compare $\sqrt{2}e^{-i\pi/4}$ (without the i) with $|z|e^{i\theta}$ to see that $|z| = \sqrt{2}$ and $\theta = -\frac{\pi}{4}$.

$x = |z|\cos\theta = \sqrt{2}\cos\left(-\frac{\pi}{4}\right) = \sqrt{2}\left(\frac{\sqrt{2}}{2}\right) = \frac{2}{2} = 1$

$y = |z|\sin\theta = \sqrt{2}\sin\left(-\frac{\pi}{4}\right) = \sqrt{2}\left(-\frac{\sqrt{2}}{2}\right) = -\frac{2}{2} = -1$

This gives the Cartesian form $1 - i$, but remember that we saved an overall i for later, which makes the final answer $i(1 - i) = i - i^2 = i - (-1) = i + 1 = 1 + i$.

Note: $-\frac{\pi}{4} = -\frac{\pi}{4}\frac{180°}{\pi} = -\frac{180°}{4} = -45°$.

Answer Key

Chapter 12

(1) $\cos(i\pi) = \cosh(\pi) = \frac{e^{\pi}+e^{-\pi}}{2} \approx \frac{23.1407+0.0432}{2} \approx 11.59$

Note: Most of the problems in this chapter require using **radians** (and **not** degrees).

(2) $\sin(i) = i \sinh 1 = i \frac{e^1-e^{-1}}{2} \approx i \frac{2.7183-0.3679}{2} \approx 1.18i$

Note: Compare $\sin(i)$ with $\sin(i\theta)$ to see that $\theta = 1$ radian.

(3) $\cos\left(-\frac{\pi}{6}i\right) = \cosh\left(-\frac{\pi}{6}\right) = \frac{e^{-\pi/6}+e^{-\pi/6}}{2} \approx \frac{0.5924+1.6881}{2} \approx 1.14$

(4) $\sin(i \ln 2) = i \sinh(\ln 2) = i \frac{e^{\ln 2}-e^{-\ln 2}}{2} = \frac{i}{2}\left[e^{\ln 2} - e^{-\ln 2}\right] = \frac{i}{2}\left[e^{\ln 2} - e^{\ln(1/2)}\right] = \frac{i}{2}\left(2 - \frac{1}{2}\right) = \frac{i}{2}\left(\frac{3}{2}\right) = \frac{3}{4}i$

Notes: $e^{-\ln 2} = e^{\ln(1/2)}$ because $-\ln t = \ln\left(\frac{1}{t}\right)$. We also used the identity $e^{\ln u} = u$.

(5) $\cosh 0 = \frac{e^0+e^{-0}}{2} = \frac{e^0+e^0}{2} = \frac{1+1}{2} = \frac{2}{2} = 1$

Notes: Of course, $-0 = 0$. This and the next three problems have nothing to do with complex numbers. They just use the definitions of hyperbolic cosine and sine.

(6) $\sinh 0 = \frac{e^0-e^{-0}}{2} = \frac{e^0-e^0}{2} = \frac{1-1}{2} = \frac{0}{2} = 0$

(7) $\cosh 1 = \frac{e^1+e^{-1}}{2} \approx \frac{2.7183+0.3679}{2} \approx 1.54$

(8) $\sinh 1 = \frac{e^1-e^{-1}}{2} \approx \frac{2.7183-0.3679}{2} \approx 1.18$

(9) $\cosh\left(\frac{\pi}{3}i\right) = \cos\left(\frac{\pi}{3}\right) = \frac{1}{2}$ Note: $\frac{\pi}{3}$ radians $= \frac{\pi}{3}\frac{180°}{\pi} = 60°$.

(10) $\sinh\left(\frac{\pi}{2}i\right) = i \sin\left(\frac{\pi}{2}\right) = i(1) = i$ Note: $\frac{\pi}{2}$ radians $= \frac{\pi}{2}\frac{180°}{\pi} = 90°$.

(11) $\cosh(i\pi) = \cos(\pi) = -1$ Note: π radians $= 180°$.

(12) $\sinh\left(\frac{5\pi}{6}i\right) = i \sin\left(\frac{5\pi}{6}\right) = i\left(\frac{1}{2}\right) = \frac{i}{2}$ Note: $\frac{5\pi}{6}$ radians $= \frac{5\pi}{6}\frac{180°}{\pi} = \frac{900°}{6} = 150°$.

(13) $\cosh\left(\frac{7\pi}{6}i\right) = \cos\left(\frac{7\pi}{6}\right) = -\frac{\sqrt{3}}{2}$ Note: $\frac{7\pi}{6}$ radians $= \frac{7\pi}{6}\frac{180°}{\pi} = 210°$.

(14) $\sinh\left(-\frac{\pi}{4}i\right) = i \sin\left(-\frac{\pi}{4}\right) = i\left(-\frac{\sqrt{2}}{2}\right) = -\frac{\sqrt{2}}{2}i$ Note: $-\frac{\pi}{4}$ radians $= -\frac{\pi}{4}\frac{180°}{\pi} = -45°$.

Alternate answer: $-\frac{i}{\sqrt{2}}$ (since $\frac{1}{\sqrt{2}} = \frac{1}{\sqrt{2}}\frac{\sqrt{2}}{\sqrt{2}} = \frac{\sqrt{2}}{2}$)

(15) $\tan(i) = \frac{\sin(i)}{\cos(i)} = \frac{i \sinh 1}{\cosh 1} = i\frac{e^1-e^{-1}}{e^1+e^{-1}} \approx i\frac{2.7183-0.3679}{2.7183+0.3679} \approx i\frac{2.350}{3.086} \approx 0.762i$

182

Note: The 2's cancel out in $\frac{e^1-e^{-1}}{2} \div \frac{e^1+e^{-1}}{2}$ since $\frac{a}{2} \div \frac{b}{2} = \frac{a}{2} \times \frac{2}{b} = \frac{2a}{2b} = \frac{a}{b}$.

(16) $\tan(-i \ln 2) = \frac{\sin(-i \ln 2)}{\cos(-i \ln 2)} = \frac{i \sinh(-\ln 2)}{\cosh(-\ln 2)} = i \frac{e^{-\ln 2} - e^{\ln 2}}{e^{-\ln 2} + e^{\ln 2}} = i \frac{e^{\ln(1/2)} - e^{\ln 2}}{e^{\ln(1/2)} + e^{\ln 2}} = i \frac{\frac{1}{2} - 2}{\frac{1}{2} + 2} = i \frac{\frac{1}{2} - \frac{4}{2}}{\frac{1}{2} + \frac{4}{2}} = -i \frac{3/2}{5/2} = -\frac{3}{5}i$

Notes: The 2's cancel out in $\frac{e^{-\ln 2} - e^{\ln 2}}{2} \div \frac{e^{-\ln 2} + e^{\ln 2}}{2}$ since $\frac{a}{2} \div \frac{b}{2} = \frac{a}{2} \times \frac{2}{b} = \frac{2a}{2b} = \frac{a}{b}$. $e^{-\ln 2} = e^{\ln(1/2)}$ because $-\ln t = \ln\left(\frac{1}{t}\right)$. We also used the identity $e^{\ln u} = u$.

(17) $\tanh\left(\frac{\pi}{6}i\right) = \frac{\sinh(i\pi/6)}{\cosh(i\pi/6)} = \frac{i \sin(\pi/6)}{\cos(\pi/6)} = \frac{i/2}{\sqrt{3}/2} = \frac{i}{\sqrt{3}} = \frac{\sqrt{3}}{3}i$

Notes: $\frac{\pi}{6}$ radians $= \frac{\pi}{6} \frac{180°}{\pi} = 30°$. $\frac{1/2}{\sqrt{3}/2} = \frac{1}{2} \div \frac{\sqrt{3}}{2} = \frac{1}{2} \times \frac{2}{\sqrt{3}} = \frac{2}{2\sqrt{3}} = \frac{1}{\sqrt{3}} = \frac{1}{\sqrt{3}} \frac{\sqrt{3}}{\sqrt{3}} = \frac{\sqrt{3}}{3}$. Recall that the way to divide by a fraction is to multiply by its reciprocal.

(18) $\tanh\left(-\frac{\pi}{4}i\right) = \frac{\sinh(-i\pi/4)}{\cosh(-i\pi/4)} = \frac{i \sin(-\pi/4)}{\cos(-\pi/4)} = \frac{-i\sqrt{2}/2}{\sqrt{2}/2} = -i$

Notes: $\frac{\pi}{4}$ radians $= \frac{\pi}{4} \frac{180°}{\pi} = 45°$. $\frac{\sqrt{2}/2}{\sqrt{2}/2} = \frac{\sqrt{2}}{2} \div \frac{\sqrt{2}}{2} = \frac{\sqrt{2}}{2} \times \frac{2}{\sqrt{2}} = \frac{2\sqrt{2}}{2\sqrt{2}} = 1$. Recall that the way to divide by a fraction is to multiply by its reciprocal.

(19) Apply $\sin(\theta + i\varphi) = \sin\theta \cosh\varphi + i\cos\theta \sinh\varphi$ with $\theta = \frac{\pi}{2}$ and $\varphi = \frac{1}{2}$.

$\sin\left(\frac{\pi}{2} + \frac{i}{2}\right) = \sin\frac{\pi}{2} \cosh\frac{1}{2} + i\cos\frac{\pi}{2} \sinh\frac{1}{2} = (1)\frac{e^{1/2} + e^{-1/2}}{2} + i(0)\frac{e^{1/2} - e^{-1/2}}{2}$

$\approx \frac{1.6487 + 0.6065}{2} + 0 \approx 1.13$ Note: $\frac{\pi}{2}$ radians $= \frac{\pi}{2} \frac{180°}{\pi} = 90°$.

(20) Apply $\cos(\theta + i\varphi) = \cos\theta \cosh\varphi - i\sin\theta \sinh\varphi$ with $\theta = \frac{\pi}{3}$ and $\varphi = -\ln 3$.

$\cos\left(\frac{\pi}{3} - i\ln 3\right) = \cos\frac{\pi}{3} \cosh(-\ln 3) - i\sin\frac{\pi}{3} \sinh(-\ln 3)$

$= \left(\frac{1}{2}\right)\frac{e^{-\ln 3} + e^{\ln 3}}{2} - i\left(\frac{\sqrt{3}}{2}\right)\frac{e^{-\ln 3} - e^{\ln 3}}{2} = \left(\frac{1}{4}\right)\left[e^{\ln(1/3)} + e^{\ln 3}\right] - i\left(\frac{\sqrt{3}}{4}\right)\left[e^{\ln(1/3)} - e^{\ln 3}\right]$

$= \left(\frac{1}{4}\right)\left(\frac{1}{3} + 3\right) - i\left(\frac{\sqrt{3}}{4}\right)\left(\frac{1}{3} - 3\right) = \frac{1}{4}\left(\frac{1}{3} + \frac{9}{3}\right) - i\left(\frac{\sqrt{3}}{4}\right)\left(\frac{1}{3} - \frac{9}{3}\right) = \frac{1}{4}\left(\frac{10}{3}\right) - i\left(\frac{\sqrt{3}}{4}\right)\left(-\frac{8}{3}\right)$

$= \frac{10}{12} + \frac{2\sqrt{3}}{3}i = \frac{5}{6} + \frac{2\sqrt{3}}{3}i$ Alternate answer: $\frac{5}{6} + \frac{2}{\sqrt{3}}i$ (since $\frac{1}{\sqrt{3}} = \frac{1}{\sqrt{3}}\frac{\sqrt{3}}{\sqrt{3}} = \frac{\sqrt{3}}{3}$)

Notes: $\frac{\pi}{3}$ radians $= \frac{\pi}{3} \frac{180°}{\pi} = 60°$. $e^{-\ln 3} = e^{\ln(1/3)}$ because $-\ln t = \ln\left(\frac{1}{t}\right)$. We also used the identity $e^{\ln u} = u$.

(21) Note that $\frac{i}{3} - \frac{\pi}{6} = -\frac{\pi}{6} + \frac{i}{3}$. Apply $\sin(\theta + i\varphi) = \sin\theta \cosh\varphi + i\cos\theta \sinh\varphi$ with

183

Answer Key

$\theta = -\frac{\pi}{6}$ and $\varphi = \frac{1}{3}$.

$\sin\left(\frac{i}{3} - \frac{\pi}{6}\right) = \sin\left(-\frac{\pi}{6}\right)\cosh\frac{1}{3} + i\cos\left(-\frac{\pi}{6}\right)\sinh\frac{1}{3} = \left(-\frac{1}{2}\right)\frac{e^{1/3}+e^{-1/3}}{2} + i\left(\frac{\sqrt{3}}{2}\right)\frac{e^{1/3}-e^{-1/3}}{2}$

$\approx -\frac{1.3956+0.7165}{4} + i\sqrt{3}\,\frac{1.3956-0.7165}{4} \approx -0.53 + 0.17i\sqrt{3} \approx -0.53 + 0.29i$

Note: $-\frac{\pi}{6}$ radians $= -\frac{\pi}{6}\frac{180°}{\pi} = -30°$.

(22) Apply $\cosh(\theta + i\varphi) = \cosh\theta\cos\varphi + i\sinh\theta\sin\varphi$ with $\theta = \frac{1}{2}$ and $\varphi = \pi$.

$\cosh\left(\frac{1}{2} + i\pi\right) = \cosh\frac{1}{2}\cos\pi + i\sinh\frac{1}{2}\sin\pi = \frac{e^{1/2}+e^{-1/2}}{2}(-1) + i\frac{e^{1/2}-e^{-1/2}}{2}(0)$

$\approx -\frac{1.6487+0.6065}{2} + 0 \approx -1.13$ Note: π radians $= 180°$.

(23) Apply $\sinh(\theta + i\varphi) = \sinh\theta\cos\varphi + i\cosh\theta\sin\varphi$ with $\theta = \ln 4$ and $\varphi = \frac{2\pi}{3}$.

$\sinh\left(\ln 4 + \frac{2\pi}{3}i\right) = \sinh(\ln 4)\cos\frac{2\pi}{3} + i\cosh(\ln 4)\sin\frac{2\pi}{3}$

$= \frac{e^{\ln 4}-e^{-\ln 4}}{2}\left(-\frac{1}{2}\right) + i\frac{e^{\ln 4}+e^{-\ln 4}}{2}\left(\frac{\sqrt{3}}{2}\right) = -\frac{1}{4}\left[e^{\ln 4} - e^{\ln(1/4)}\right] + \frac{i\sqrt{3}}{4}\left[e^{\ln 4} + e^{\ln(1/4)}\right]$

$= -\frac{1}{4}\left(4 - \frac{1}{4}\right) + \frac{i\sqrt{3}}{4}\left(4 + \frac{1}{4}\right) = -\frac{1}{4}\left(\frac{16}{4} - \frac{1}{4}\right) + \frac{i\sqrt{3}}{4}\left(\frac{16}{4} + \frac{1}{4}\right) = -\frac{1}{4}\frac{15}{4} + \frac{i\sqrt{3}}{4}\frac{17}{4} = -\frac{15}{16} + \frac{17i\sqrt{3}}{16}$

Notes: $\frac{2\pi}{3}$ radians $= \frac{2\pi}{3}\frac{180°}{\pi} = 120°$. $e^{-\ln 4} = e^{\ln(1/4)}$ because $-\ln t = \ln\left(\frac{1}{t}\right)$. We also used the identity $e^{\ln u} = u$.

(24) Note that $\frac{\pi}{2}i - e = -e + \frac{\pi}{2}i$. Apply $\cosh(\theta + i\varphi) = \cosh\theta\cos\varphi + i\sinh\theta\sin\varphi$ with $\theta = -e$ and $\varphi = \frac{\pi}{2}$.

$\cosh\left(\frac{\pi}{2}i - e\right) = \cosh(-e)\cos\frac{\pi}{2} + i\sinh(-e)\sin\frac{\pi}{2} = \frac{e^{-e}+e^{-(-e)}}{2}(0) + i\frac{e^{-e}-e^{-(-e)}}{2}(1)$

$= i\frac{e^{-e}-e^{e}}{2} \approx i\frac{0.0660-15.1543}{2} \approx -7.54i$ Note: $\frac{\pi}{2}$ radians $= \frac{\pi}{2}\frac{180°}{\pi} = 90°$.

(25) See Example 5. The $^{-1}$ does **not** mean reciprocal; it's an inverse.

Use the formula $\cosh^{-1}(iz) = \pm i\cos^{-1}z$ with $z = \frac{1}{2}$. This gives $\cosh^{-1}\left(\frac{i}{2}\right) = \pm i\cos^{-1}\left(\frac{1}{2}\right)$. Of the possible answers, the ones that make hyperbolic cosine equal positive one-half are $\frac{\pi}{3}i + 2\pi ik$ or $-\frac{\pi}{3}i + 2\pi ik$ (equivalent to $\frac{5\pi}{3} + 2\pi ik$), which can be condensed down to $\pm\frac{\pi}{3}i + 2\pi ik$.

Check the answer: $\cosh\left(\pm\frac{\pi}{3}i\right) = \cos\left(\pm\frac{\pi}{3}\right) = \frac{1}{2}$.

Complex Numbers Essentials Math Workbook with Answers

(26) See Example 5. The $^{-1}$ does **not** mean reciprocal; it's an inverse.
Use the formula $\sinh^{-1}(iz) = i\sin^{-1} z$ with $z = \frac{\sqrt{2}}{2}$. This gives $\sinh^{-1}\left(\frac{\sqrt{2}}{2}i\right) =$
$i\sin^{-1}\frac{\sqrt{2}}{2}$. The answers are $\frac{\pi}{4}i - 2\pi i k$ or $\frac{3\pi}{4}i - 2\pi i k$.
Check the answer: $\sinh\left(\frac{\pi}{4}i\right) = \sinh\left(\frac{3\pi}{4}i\right) = i\sin\left(\frac{\pi}{4}\right) = i\sin\left(\frac{3\pi}{4}\right) = \frac{\sqrt{2}}{2}i$.

(27) Follow Example 6. Use the formula $\ln z = \ln(|z|e^{i\theta}) = i\theta + 2\pi i k + \ln|z|$.
Compare $z = \frac{\sqrt{2}}{2} + \frac{\sqrt{2}}{2}i$ with $z = x + iy$ to see that $x = \frac{\sqrt{2}}{2}$ and $y = \frac{\sqrt{2}}{2}$, such that $|z| =$
$\sqrt{x^2 + y^2} = \sqrt{\left(\frac{\sqrt{2}}{2}\right)^2 + \left(\frac{\sqrt{2}}{2}\right)^2} = \sqrt{\frac{2}{4} + \frac{2}{4}} = \sqrt{\frac{4}{4}} = \sqrt{1} = 1$ and $\theta = \tan^{-1}\left(\frac{y}{x}\right) =$
$\tan^{-1} 1 = \frac{\pi}{4}$ (recall Chapter 7). This gives $\ln\left(\frac{\sqrt{2}}{2} + \frac{\sqrt{2}}{2}i\right) = \ln(1 e^{i\pi/4}) = \frac{i\pi}{4} + 2\pi i k +$
$\ln 1 = \frac{i\pi}{4} + 2\pi i k + 0 = \frac{i\pi}{4} + 2\pi i k$.
Check the answer: $e^{i\pi/4} = \cos\frac{\pi}{4} + i\sin\frac{\pi}{4} = \frac{\sqrt{2}}{2} + \frac{\sqrt{2}}{2}i$ (using Euler's formula). If we
take the natural logarithm of both sides of this equation, we get $\frac{\pi}{4}i = \ln\left(\frac{\sqrt{2}}{2} + \frac{\sqrt{2}}{2}i\right)$.
The $2\pi i k$ doesn't affect anything since $e^{2\pi i k} = 1$ and $e^{i\pi/4 + 2\pi i k} = e^{i\pi/4}e^{2\pi i k}$.

(28) $\sin\theta = \frac{e^{i\theta} - e^{-i\theta}}{2i} \to \sin(i\theta) = \frac{e^{i(i\theta)} - e^{-i(i\theta)}}{2i} = \frac{e^{i^2\theta} - e^{-i^2\theta}}{2i} = \frac{e^{-\theta} - e^{-(-1)\theta}}{2i} = \frac{e^{-\theta} - e^{\theta}}{2i}$
$\sin(i\theta) = \frac{e^{-\theta} - e^{\theta}}{2}\frac{1}{i} = \frac{e^{-\theta} - e^{\theta}}{2}\frac{1\,i}{i\,i} = \frac{e^{-\theta} - e^{\theta}}{2}\frac{i}{i^2} = \frac{e^{-\theta} - e^{\theta}}{2}\frac{i}{-1} = i\frac{-1(e^{-\theta} - e^{\theta})}{2} = i\frac{-e^{-\theta} - (-e^{\theta})}{2}$
$\sin(i\theta) = i\frac{-e^{-\theta} + e^{\theta}}{2} = i\frac{e^{\theta} - e^{-\theta}}{2} = i\sinh\theta$
Note: In the last step, we used the fact that $-a + b = b - a$.

(29) $\cosh\theta = \frac{e^{\theta} + e^{-\theta}}{2} \to \cosh(i\theta) = \frac{e^{(i\theta)} + e^{-(i\theta)}}{2} = \frac{e^{i\theta} + e^{-i\theta}}{2} = \cos\theta$

(30) $\sinh\theta = \frac{e^{\theta} - e^{-\theta}}{2} \to \sinh(i\theta) = \frac{e^{(i\theta)} - e^{-(i\theta)}}{2} = \frac{e^{i\theta} - e^{-i\theta}}{2}$
$i\sin\theta = i\frac{e^{i\theta} - e^{-i\theta}}{2i} = \frac{e^{i\theta} - e^{-i\theta}}{2}$ Since $\sinh(i\theta) = \frac{e^{i\theta} - e^{-i\theta}}{2}$ and $i\sin\theta = \frac{e^{i\theta} - e^{-i\theta}}{2}$, it follows
that $\sinh(i\theta) = i\sin\theta$. Notes: We used $\sinh\theta = \frac{e^{\theta} - e^{-\theta}}{2}$ to show that $\sinh(i\theta) = \frac{e^{i\theta} - e^{-i\theta}}{2}$ and we separately used $\sin\theta = \frac{e^{i\theta} - e^{-i\theta}}{2i}$ to show that $i\sin\theta = \frac{e^{i\theta} - e^{-i\theta}}{2}$. Then
we applied the **transitive property** of algebra: If $a = c$ and $b = c$, then $a = b$.

Answer Key

(31) Begin with the usual angle sum identity for sine from trigonometry:
$\sin(\theta + \beta) = \sin\theta \cos\beta + \cos\theta \sin\beta$. Now substitute $\beta = i\varphi$.
$\sin(\theta + i\varphi) = \sin\theta \cos(i\varphi) + \cos\theta \sin(i\varphi) = \sin\theta \cosh\varphi + i\cos\theta \sinh\varphi$
Notes: We used the identities $\cos(i\varphi) = \cosh\varphi$ and $\sin(i\varphi) = i\sinh\varphi$.

(32) If you happen to know (or if you look up) the angle sum formula for the usual real hyperbolic cosine, you can use it and solve this problem similar to Problem 31. If you don't know that 'magic formula,' it doesn't really matter. We'll show you that you could solve this problem without knowing it. We'll take the definition of hyperbolic cosine and replace the angle with $\theta + i\varphi$.

$\cosh\theta = \dfrac{e^\theta + e^{-\theta}}{2} \to \cosh(\theta + i\varphi) = \dfrac{e^{\theta+i\varphi} + e^{-\theta-i\varphi}}{2}$

Next, we'll work out $\cosh\theta \cos\varphi + i\sinh\theta \sin\varphi$:

$\cosh\theta \cos\varphi + i\sinh\theta \sin\varphi = \dfrac{e^\theta + e^{-\theta}}{2}\dfrac{e^{i\varphi}+e^{-i\varphi}}{2} + i\dfrac{e^\theta - e^{-\theta}}{2}\dfrac{e^{i\varphi}-e^{-i\varphi}}{2i}$ (note: the i's cancel)

$\cosh\theta \cos\varphi + i\sinh\theta \sin\varphi = \dfrac{e^\theta e^{i\varphi} + e^\theta e^{-i\varphi} + e^{-\theta}e^{i\varphi} + e^{-\theta}e^{-i\varphi}}{4} + \dfrac{e^\theta e^{i\varphi} - e^\theta e^{-i\varphi} - e^{-\theta}e^{i\varphi} + e^{-\theta}e^{-i\varphi}}{4}$

$\cosh\theta \cos\varphi + i\sinh\theta \sin\varphi = \dfrac{e^{\theta+i\varphi} + e^{\theta-i\varphi} + e^{-\theta+i\varphi} + e^{-\theta-i\varphi} + e^{\theta+i\varphi} - e^{\theta-i\varphi} - e^{-\theta+i\varphi} + e^{-\theta-i\varphi}}{4}$

$\cosh\theta \cos\varphi + i\sinh\theta \sin\varphi = \dfrac{2e^{\theta+i\varphi} + 2e^{-\theta-i\varphi}}{4} = \dfrac{e^{\theta+i\varphi} + e^{-\theta-i\varphi}}{2}$

This shows that $\cosh(\theta + i\varphi)$ and $\cosh\theta \cos\varphi + i\sinh\theta \sin\varphi$ are both equal to $\dfrac{e^{\theta+i\varphi} + e^{-\theta-i\varphi}}{2}$. It follows that $\cosh(\theta + i\varphi) = \cosh\theta \cos\varphi + i\sinh\theta \sin\varphi$.

(33) We'll solve this the way that we solved Problem 32, though if you know (or look up) the usual angle sum identity for hyperbolic sine, you could use that instead.

$\sinh\theta = \dfrac{e^\theta - e^{-\theta}}{2} \to \sinh(\theta + i\varphi) = \dfrac{e^{\theta+i\varphi} - e^{-\theta-i\varphi}}{2}$

$\sinh\theta \cos\varphi + i\cosh\theta \sin\varphi = \dfrac{e^\theta - e^{-\theta}}{2}\dfrac{e^{i\varphi}+e^{-i\varphi}}{2} + i\dfrac{e^\theta + e^{-\theta}}{2}\dfrac{e^{i\varphi}-e^{-i\varphi}}{2i}$ (note: the i's cancel)

$\sinh\theta \cos\varphi + i\cosh\theta \sin\varphi = \dfrac{e^\theta e^{i\varphi} + e^\theta e^{-i\varphi} - e^{-\theta}e^{i\varphi} - e^{-\theta}e^{-i\varphi}}{4} + \dfrac{e^\theta e^{i\varphi} - e^\theta e^{-i\varphi} + e^{-\theta}e^{i\varphi} - e^{-\theta}e^{-i\varphi}}{4}$

$\sinh\theta \cos\varphi + i\cosh\theta \sin\varphi = \dfrac{e^{\theta+i\varphi} + e^{\theta-i\varphi} - e^{-\theta+i\varphi} - e^{-\theta-i\varphi} + e^{\theta+i\varphi} - e^{\theta-i\varphi} + e^{-\theta+i\varphi} - e^{-\theta-i\varphi}}{4}$

$\sinh\theta \cos\varphi + i\cosh\theta \sin\varphi = \dfrac{2e^{\theta+i\varphi} - 2e^{-\theta-i\varphi}}{4} = \dfrac{e^{\theta+i\varphi} - e^{-\theta-i\varphi}}{2}$

This shows that $\sinh(\theta + i\varphi)$ and $\sinh\theta \cos\varphi + i\cosh\theta \sin\varphi$ are both equal to $\dfrac{e^{\theta+i\varphi} - e^{-\theta-i\varphi}}{2}$. It follows that $\sinh(\theta + i\varphi) = \sinh\theta \cos\varphi + i\cosh\theta \sin\varphi$.

Chapter 13

(1) Compare $65 - 72i$ with $z = x + iy$ to see that $x = 65$ and $y = -72$.

$$|z| = \sqrt{x^2 + y^2} = \sqrt{65^2 + (-72)^2} = \sqrt{4225 + 5184} = \sqrt{9409} = 97$$

$$\theta = \tan^{-1}\left(\frac{y}{x}\right) = \tan^{-1}\left(\frac{-72}{65}\right) \approx -0.8364 \text{ rad (Quad. IV)}$$

Make sure your calculator is in **radians** mode. (If you get -47.92, that's degrees.)

$$97^{1/2} e^{-0.8364 i/2} = 97^{1/2} e^{-0.4182 i}$$

$$97^{1/2} e^{-0.4182 i} = 97^{1/2} \cos(-0.4182) + i 97^{1/2} \sin(-0.4182) \approx 9 - 4i$$

When $n = 2$, the second solution is the negative of the first solution: $-9 + 4i$.

Check: $(9 - 4i)(9 - 4i) = 9(9) + 9(-4i) - 4i(9) - 4i(-4i) = 81 - 36i - 36i + 16i^2$
$= 81 - 72i - 16 = 65 - 72i$

$(-9 + 4i)(-9 + 4i) = -9(-9) - 9(4i) + 4i(-9) + 4i(4i) = 81 - 36i - 36i + 16i^2$
$= 81 - 72i - 16 = 65 - 72i$

(2) Compare $-3i$ with $z = x + iy$ to see that $x = 0$ and $y = -3$.

$$|z| = \sqrt{x^2 + y^2} = \sqrt{0^2 + (-3)^2} = \sqrt{0 + 9} = \sqrt{9} = 3$$

$$\theta = \tan^{-1}\left(\frac{y}{x}\right) = \tan^{-1}\left(\frac{-3}{0}\right) = \frac{3\pi}{2} \text{ rad or } -\frac{\pi}{2} \text{ rad (on the neg. } y\text{-axis)}$$

Since $x = 0$ and $y = -3$, the point lies on the $-y$-axis, which has an angle of $270°$ (equivalent to $\frac{3\pi}{2}$ rad or $-\frac{\pi}{2}$ rad). See the solution to Chapter 7, Exercise 9.

$$3^{1/2} e^{(3i\pi/2)/2} = 3^{1/2} e^{3i\pi/4}$$

$$3^{1/2} e^{3i\pi/4} = 3^{1/2} \cos\frac{3i\pi}{4} + i 3^{1/2} \sin\frac{3i\pi}{4} = \sqrt{3}\left(-\frac{\sqrt{2}}{2}\right) + i\sqrt{3}\left(\frac{\sqrt{2}}{2}\right) = -\frac{\sqrt{6}}{2} + \frac{\sqrt{6}}{2} i$$

When $n = 2$, the second solution is the negative of the first solution: $\frac{\sqrt{6}}{2} - \frac{\sqrt{6}}{2} i$.

Alternate answer: $-\sqrt{3/2} + i\sqrt{3/2}$ since $\frac{\sqrt{3}}{\sqrt{2}} = \frac{\sqrt{3}\sqrt{2}}{\sqrt{2}\sqrt{2}} = \frac{\sqrt{6}}{2}$.

Check: $\left(-\frac{\sqrt{6}}{2} + \frac{\sqrt{6}}{2} i\right)\left(-\frac{\sqrt{6}}{2} + \frac{\sqrt{6}}{2} i\right) = -\frac{\sqrt{6}}{2}\left(-\frac{\sqrt{6}}{2}\right) - \frac{\sqrt{6}}{2}\left(\frac{\sqrt{6}}{2} i\right) + \frac{\sqrt{6}}{2} i\left(-\frac{\sqrt{6}}{2}\right) + \frac{\sqrt{6}}{2} i\left(\frac{\sqrt{6}}{2} i\right)$

$= \frac{6}{4} - \frac{6}{4} i - \frac{6}{4} i + \frac{6}{4} i^2 = \frac{3}{2} - \frac{3}{2} i - \frac{3}{2} i + \frac{3}{2}(-1) = 0 - 3i = -3i$

$\left(\frac{\sqrt{6}}{2} - \frac{\sqrt{6}}{2} i\right)\left(\frac{\sqrt{6}}{2} - \frac{\sqrt{6}}{2} i\right) = \frac{\sqrt{6}}{2}\left(\frac{\sqrt{6}}{2}\right) + \frac{\sqrt{6}}{2}\left(-\frac{\sqrt{6}}{2} i\right) - \frac{\sqrt{6}}{2} i\left(\frac{\sqrt{6}}{2}\right) - \frac{\sqrt{6}}{2} i\left(-\frac{\sqrt{6}}{2} i\right)$

$= \frac{6}{4} - \frac{6}{4} i - \frac{6}{4} i + \frac{6}{4} i^2 = \frac{3}{2} - \frac{3}{2} i - \frac{3}{2} i + \frac{3}{2}(-1) = 0 - 3i = -3i$

Answer Key

(3) Note that $i - 1 = -1 + i$. Compare $-1 + i$ with $z = x + iy$ to see that $x = -1$ and $y = 1$.

$$|z| = \sqrt{x^2 + y^2} = \sqrt{(-1)^2 + 1^2} = \sqrt{1 + 1} = \sqrt{2}$$

$$\theta = \tan^{-1}\left(\frac{y}{x}\right) = \tan^{-1}\left(\frac{1}{-1}\right) = -\frac{\pi}{4} + \pi = \frac{3\pi}{4} \text{ rad (Quad. II)}$$

Notes: Add π rad to the calculator's answer for inverse tangent if $x < 0$ (Chapter 7). Make sure your calculator is in **radians** mode.

$$\sqrt{2}^{1/3} e^{(3i\pi/4)/3} = 2^{1/6} e^{i\pi/4}$$

Note: Since $|z| = \sqrt{2}$, it follows that $|z|^{1/3} = \sqrt{2}^{1/3} = \left(2^{1/2}\right)^{1/3} = 2^{1/6}$ according to the rule $(a^b)^c = a^{bc}$ since $\frac{1}{2}\left(\frac{1}{3}\right) = \frac{1}{6}$. Wondering why there is a 1/4 instead of a 1/3 in $e^{i\pi/4}$? The angle was originally $\frac{3\pi}{4}$. When we divided by 3, the 3 canceled out.

$$2^{1/6} e^{i\pi/4} = 2^{1/6} \cos\frac{\pi}{4} + i 2^{1/6} \sin\frac{\pi}{4} = 2^{1/6}\left(\frac{\sqrt{2}}{2}\right) + i 2^{1/6}\left(\frac{\sqrt{2}}{2}\right)$$

$$= \frac{2^{1/6} 2^{1/2}}{2} + \frac{2^{1/6} 2^{1/2}}{2} i = \frac{2^{2/3}}{2} + \frac{2^{2/3}}{2} i \text{ Alternate answer: } \frac{1}{2^{1/3}} + \frac{i}{2^{1/3}}$$

Notes: We used the rule $x^m x^n = x^{m+n}$ to get $2^{1/6} 2^{1/2} = 2^{2/3}$ and the rule $\frac{x^m}{x^n} = x^{m-n}$ to get $\frac{2^{2/3}}{2} = 2^{2/3-1} = 2^{-1/3} = \frac{1}{2^{1/3}}$. The answer $\frac{2^{2/3}}{2} + \frac{2^{2/3}}{2} i$ is preferred over $\frac{1}{2^{1/3}} + \frac{i}{2^{1/3}}$ because it has a rational denominator. $\left(\frac{1}{6} + \frac{1}{2} = \frac{1}{6} + \frac{3}{6} = \frac{4}{6} = \frac{2}{3}.\right)$

The first solution is $x_1 + iy_1 = \frac{2^{2/3}}{2} + \frac{2^{2/3}}{2} i$. The second and third solutions are:

$$x_2 = x_1 \cos\left(\frac{2\pi}{3}\right) - y_1 \sin\left(\frac{2\pi}{3}\right) = \frac{2^{2/3}}{2}\left(-\frac{1}{2}\right) - \frac{2^{2/3}}{2}\left(\frac{\sqrt{3}}{2}\right) = -\frac{2^{2/3}}{4} - \frac{2^{2/3}\sqrt{3}}{4}$$

$$y_2 = x_1 \sin\left(\frac{2\pi}{3}\right) + y_1 \cos\left(\frac{2\pi}{3}\right) = \frac{2^{2/3}}{2}\left(\frac{\sqrt{3}}{2}\right) + \frac{2^{2/3}}{2}\left(-\frac{1}{2}\right) = \frac{2^{2/3}\sqrt{3}}{4} - \frac{2^{2/3}}{4}$$

$$x_2 + iy_2 = -\frac{2^{2/3}}{4} - \frac{2^{2/3}\sqrt{3}}{4} + i\frac{2^{2/3}\sqrt{3}}{4} - \frac{2^{2/3}}{4}i = -\frac{2^{2/3}}{4}(1 + \sqrt{3} - i\sqrt{3} + i)$$

$$x_3 = x_1 \cos\left(\frac{4\pi}{3}\right) - y_1 \sin\left(\frac{4\pi}{3}\right) = \frac{2^{2/3}}{2}\left(-\frac{1}{2}\right) - \frac{2^{2/3}}{2}\left(-\frac{\sqrt{3}}{2}\right) = -\frac{2^{2/3}}{4} + \frac{2^{2/3}\sqrt{3}}{4}$$

$$y_3 = x_1 \sin\left(\frac{4\pi}{3}\right) + y_1 \cos\left(\frac{4\pi}{3}\right) = \frac{2^{2/3}}{2}\left(-\frac{\sqrt{3}}{2}\right) + \frac{2^{2/3}}{2}\left(-\frac{1}{2}\right) = -\frac{2^{2/3}\sqrt{3}}{4} - \frac{2^{2/3}}{4}$$

$$x_3 + iy_3 = -\frac{2^{2/3}}{4} + \frac{2^{2/3}\sqrt{3}}{4} - i\frac{2^{2/3}\sqrt{3}}{4} - \frac{2^{2/3}}{4}i = -\frac{2^{2/3}}{4}(1 - \sqrt{3} + i\sqrt{3} + i)$$

Check: $\left(\frac{2^{2/3}}{2} + \frac{2^{2/3}}{2}i\right)\left(\frac{2^{2/3}}{2} + \frac{2^{2/3}}{2}i\right) = \frac{2^{2/3}}{2}\frac{2^{2/3}}{2} + \frac{2^{2/3}}{2}i\frac{2^{2/3}}{2} + \frac{2^{2/3}}{2}i\frac{2^{2/3}}{2} + \frac{2^{2/3}}{2}i\frac{2^{2/3}}{2}i$

$= \frac{2^{4/3}}{4} + \frac{2^{4/3}}{4}i + \frac{2^{4/3}}{4}i - \frac{2^{4/3}}{4} = \frac{2^{4/3}}{2}i = 2^{4/3-1}i = 2^{1/3}i$

$\left(\frac{2^{2/3}}{2} + \frac{2^{2/3}}{2}i\right)^3 = 2^{1/3}i\left(\frac{2^{2/3}}{2} + \frac{2^{2/3}}{2}i\right) = \frac{2^1}{2}i + i^2\frac{2^1}{2} = i + i^2 = i - 1$

$\left(-\frac{2^{2/3}}{4}\right)(1 + \sqrt{3} - i\sqrt{3} + i)\left(-\frac{2^{2/3}}{4}\right)(1 + \sqrt{3} - i\sqrt{3} + i)$

$= \frac{2^{4/3}}{16}(1 + \sqrt{3} - i\sqrt{3} + i + \sqrt{3} + 3 - 3i + i\sqrt{3} - i\sqrt{3} - 3i + 3i^2 - i^2\sqrt{3} + i + i\sqrt{3} - i^2\sqrt{3} + i^2)$

$= \frac{2^{4/3}}{16}[4 + 2\sqrt{3} - 4i + 3(-1) - (-1)\sqrt{3} - (-1)\sqrt{3} + (-1)]$

$= \frac{2^{4/3}}{16}(4 + 2\sqrt{3} - 4i - 3 + \sqrt{3} + \sqrt{3} - 1) = \frac{2^{4/3}}{16}(4\sqrt{3} - 4i) = \frac{2^{4/3}}{4}(\sqrt{3} - i)$

$\frac{2^{4/3}}{4}(\sqrt{3} - i)\left(-\frac{2^{2/3}}{4}\right)(1 + \sqrt{3} - i\sqrt{3} + i) = -\frac{2^2}{16}(\sqrt{3} + 3 - 3i + i\sqrt{3} - i - i\sqrt{3} + i^2\sqrt{3} - i^2)$

$= -\frac{4}{16}[\sqrt{3} + 3 - 4i + (-1)\sqrt{3} - (-1)] = -\frac{1}{4}(3 - 4i + 1) = -\frac{1}{4}(4 - 4i) = -1 + i = i - 1$

(4) Compare $-1 + i\sqrt{3}$ with $z = x + iy$ to see that $x = -1$ and $y = \sqrt{3}$.

$|z| = \sqrt{x^2 + y^2} = \sqrt{(-1)^2 + (\sqrt{3})^2} = \sqrt{1 + 3} = \sqrt{4} = 2$

$\theta = \tan^{-1}\left(\frac{y}{x}\right) = \tan^{-1}\left(\frac{\sqrt{3}}{-1}\right) = -\frac{\pi}{3} + \pi = \frac{2\pi}{3}$ rad (Quad. II)

Notes: Add π rad to the calculator's answer for inverse tangent if $x < 0$ (Chapter 7). Make sure your calculator is in **radians** mode.

$$2^{1/4}e^{(2i\pi/3)/4} = 2^{1/4}e^{i\pi/6}$$

Notes: $\frac{2i\pi}{3}\frac{1}{4} = \frac{2i\pi}{12} = \frac{i\pi}{6}$. We applied the rule $(a^b)^c = a^{bc}$.

$2^{1/4}e^{i\pi/6} = 2^{1/4}\cos\frac{\pi}{6} + i2^{1/4}\sin\frac{\pi}{6} = 2^{1/4}\left(\frac{\sqrt{3}}{2}\right) + i2^{1/4}\left(\frac{1}{2}\right) = \frac{2^{1/4}\sqrt{3}}{2} + \frac{2^{1/4}}{2}i$

Alternate answer: $\frac{\sqrt{3}}{2^{3/4}} + \frac{i}{2^{3/4}}$ (the above form, with a rational denominator, is preferred)

The first solution is $x_1 + iy_1 = \frac{2^{1/4}\sqrt{3}}{2} + \frac{2^{1/4}}{2}i$. The 2nd, 3rd, and 4th solutions are:

$x_2 = x_1\cos\left(\frac{2\pi}{4}\right) - y_1\sin\left(\frac{2\pi}{4}\right) = \frac{2^{1/4}\sqrt{3}}{2}(0) - \frac{2^{1/4}}{2}(1) = -\frac{2^{1/4}}{2}$

$y_2 = x_1\sin\left(\frac{2\pi}{4}\right) + y_1\cos\left(\frac{2\pi}{4}\right) = \frac{2^{1/4}\sqrt{3}}{2}(1) + \frac{2^{1/4}}{2}(0) = \frac{2^{1/4}\sqrt{3}}{2}$

Answer Key

$$x_2 + iy_2 = -\frac{2^{1/4}}{2} + \frac{2^{1/4}\sqrt{3}}{2}i$$

$$x_3 = x_1 \cos\left(\frac{4\pi}{4}\right) - y_1 \sin\left(\frac{4\pi}{4}\right) = \frac{2^{1/4}\sqrt{3}}{2}(-1) - \frac{2^{1/4}}{2}(0) = -\frac{2^{1/4}\sqrt{3}}{2}$$

$$y_3 = x_1 \sin\left(\frac{4\pi}{4}\right) + y_1 \cos\left(\frac{4\pi}{4}\right) = \frac{2^{1/4}\sqrt{3}}{2}(0) + \frac{2^{1/4}}{2}(-1) = -\frac{2^{1/4}}{2}$$

$$x_3 + iy_3 = -\frac{2^{1/4}\sqrt{3}}{2} - \frac{2^{1/4}}{2}i$$

$$x_4 = x_1 \cos\left(\frac{6\pi}{4}\right) - y_1 \sin\left(\frac{6\pi}{4}\right) = \frac{2^{1/4}\sqrt{3}}{2}(0) - \frac{2^{1/4}}{2}(-1) = \frac{2^{1/4}}{2}$$

$$y_4 = x_1 \sin\left(\frac{6\pi}{4}\right) + y_1 \cos\left(\frac{6\pi}{4}\right) = \frac{2^{1/4}\sqrt{3}}{2}(-1) + \frac{2^{1/4}}{2}(0) = -\frac{2^{1/4}\sqrt{3}}{2}$$

$$x_4 + iy_4 = \frac{2^{1/4}}{2} - i\frac{2^{1/4}\sqrt{3}}{2}$$

Check: $\left(\frac{2^{1/4}\sqrt{3}}{2} + \frac{2^{1/4}}{2}i\right)\left(\frac{2^{1/4}\sqrt{3}}{2} + \frac{2^{1/4}}{2}i\right) = \frac{2^{1/4}\sqrt{3}}{2}\frac{2^{1/4}\sqrt{3}}{2} + \frac{2^{1/4}\sqrt{3}}{2}\frac{2^{1/4}}{2}i + \frac{2^{1/4}}{2}i\frac{2^{1/4}\sqrt{3}}{2} + \frac{2^{1/4}}{2}i\frac{2^{1/4}}{2}i$

$= \frac{2^{1/2}(3)}{4} + \frac{2^{1/2}\sqrt{3}}{4}i + \frac{2^{1/2}\sqrt{3}}{4}i + i^2\frac{2^{1/2}}{4} = \frac{2^{1/2}(3)}{4} + \frac{2^{1/2}\sqrt{3}}{2}i - \frac{2^{1/2}}{4} = \frac{2^{1/2}(2)}{4} + \frac{2^{1/2}\sqrt{3}}{2}i = \frac{2^{1/2}}{2} + \frac{2^{1/2}\sqrt{3}}{2}i$

$\left(\frac{2^{1/4}\sqrt{3}}{2} + i\frac{2^{1/4}}{2}\right)^4 = \left(\frac{2^{1/4}\sqrt{3}}{2} + i\frac{2^{1/4}}{2}\right)^2\left(\frac{2^{1/4}\sqrt{3}}{2} + i\frac{2^{1/4}}{2}\right)^2 = \left(\frac{2^{1/2}}{2} + i\frac{2^{1/2}\sqrt{3}}{2}\right)\left(\frac{2^{1/2}}{2} + i\frac{2^{1/2}\sqrt{3}}{2}\right)$

$= \frac{2^{1/2}}{2}\frac{2^{1/2}}{2} + \frac{2^{1/2}}{2}i\frac{2^{1/2}\sqrt{3}}{2} + i\frac{2^{1/2}\sqrt{3}}{2}\frac{2^{1/2}}{2} + i\frac{2^{1/2}\sqrt{3}}{2}i\frac{2^{1/2}\sqrt{3}}{2}$

$= \frac{2}{4} + \frac{2\sqrt{3}}{4}i + \frac{2\sqrt{3}}{4}i - \frac{2(3)}{4} = \frac{1}{2} + \frac{1\sqrt{3}}{2}i + \frac{1\sqrt{3}}{2}i - \frac{3}{2} = -1 + i\sqrt{3}$

(5) Compare $65 - 142i$ with $z = x + iy$ to see that $x = 65$ and $y = -142$.[1]

$$|z| = \sqrt{x^2 + y^2} = \sqrt{65^2 + (-142)^2} = \sqrt{4225 + 20{,}164} = \sqrt{24{,}389}$$

$$\theta = \tan^{-1}\left(\frac{y}{x}\right) = \tan^{-1}\left(\frac{-142}{65}\right) \approx -1.1415 \text{ rad (Quad. IV)}$$

Make sure your calculator is in **radians** mode. (If you get -65.4, that's degrees.)

$$\sqrt{24{,}389}^{1/3} e^{-1.1415i/3} \approx 24{,}389^{1/6} e^{-0.3805i}$$

Note: Since $|z| = \sqrt{24{,}389}$, it follows that $|z|^{1/3} = \sqrt{24{,}389}^{1/3} = \left(24{,}389^{1/2}\right)^{1/3} = 24{,}389^{1/6}$ according to the rule $(a^b)^c = a^{bc}$ since $\frac{1}{2}\left(\frac{1}{3}\right) = \frac{1}{6}$.

---

[1] With amazing arithmetic, you could factor out $29^2 = 841$ here as $\sqrt{24{,}389} = \sqrt{(841)(29)} = 29\sqrt{29}$, but since this is an intermediate answer (not a final result), it would be impressive but not important.

Complex Numbers Essentials Math Workbook with Answers

$24{,}389^{1/6}e^{-0.3805i} = 24{,}389^{1/6}\cos(-0.3805) + i\,24{,}389^{1/6}\sin(-0.3805) \approx 5 - 2i$

The first solution is $x_1 + iy_1 = 5 - 2i$. The second and third solutions are:

$$x_2 = x_1 \cos\left(\frac{2\pi}{3}\right) - y_1 \sin\left(\frac{2\pi}{3}\right) = 5\left(-\frac{1}{2}\right) - (-2)\left(\frac{\sqrt{3}}{2}\right) = -\frac{5}{2} + \sqrt{3}$$

$$y_2 = x_1 \sin\left(\frac{2\pi}{3}\right) + y_1 \cos\left(\frac{2\pi}{3}\right) = 5\left(\frac{\sqrt{3}}{2}\right) + (-2)\left(-\frac{1}{2}\right) = \frac{5\sqrt{3}}{2} + 1$$

$$x_2 + iy_2 = -\frac{5}{2} + \sqrt{3} + \frac{5\sqrt{3}}{2}i + i$$

$$x_3 = x_1 \cos\left(\frac{4\pi}{3}\right) - y_1 \sin\left(\frac{4\pi}{3}\right) = 5\left(-\frac{1}{2}\right) - (-2)\left(-\frac{\sqrt{3}}{2}\right) = -\frac{5}{2} - \sqrt{3}$$

$$y_3 = x_1 \sin\left(\frac{4\pi}{3}\right) + y_1 \cos\left(\frac{4\pi}{3}\right) = 5\left(-\frac{\sqrt{3}}{2}\right) + (-2)\left(-\frac{1}{2}\right) = -\frac{5\sqrt{3}}{2} + 1$$

$$x_3 + iy_3 = -\frac{5}{2} - \sqrt{3} - \frac{5\sqrt{3}}{2}i + i$$

Check: $(5 - 2i)(5 - 2i) = 25 - 10i - 10i + 4i^2 = 25 - 20i - 4 = 21 - 20i$

$(21 - 20i)(5 - 2i) = 105 - 42i - 100i + 40i^2 = 105 - 142i - 40 = 65 - 142i$

The other solutions are easier to check with decimal approximations. For example:

$-\frac{5}{2} + \sqrt{3} + \frac{5\sqrt{3}}{2}i + i \approx -0.768 + 5.33i$

$(-0.768 + 5.33i)(-0.768 + 5.33i) \approx 0.59 - 4.09i - 4.09i + 28.41i^2$

$= 0.59 - 8.18i - 28.41 = -27.82 - 8.18i$

$(-27.82 - 8.18i)(-0.768 + 5.33i) \approx 21.37 - 148.28i + 6.28i - 43.6i^2$

$= 21.37 - 142i + 43.6 \approx 65 - 142i$

(6) Compare i with $z = x + iy$ to see that $x = 0$ and $y = 1$.

$$|z| = \sqrt{x^2 + y^2} = \sqrt{0^2 + 1^2} = \sqrt{0 + 1} = \sqrt{1} = 1$$

$$\theta = \tan^{-1}\left(\frac{y}{x}\right) = \tan^{-1}\left(\frac{1}{0}\right) = \frac{\pi}{2} \text{ rad (on the } +y\text{-axis)}$$

Since $x = 0$ and $y = 1$, the point lies on the $+y$-axis, which has an angle of $90°$ (equivalent to $\frac{\pi}{2}$ rad). See the solution to Chapter 7, Exercise 4.

$$1^{1/3}e^{(i\pi/2)/3} = 1e^{i\pi/6} = e^{i\pi/6}$$

Note: $e^{(i\pi/2)/3} = e^{i\pi/6}$ according to the rule $(a^b)^c = a^{bc}$ since $\frac{1}{2}\left(\frac{1}{3}\right) = \frac{1}{6}$.

Answer Key

$$e^{i\pi/6} = \cos\frac{\pi}{6} + i\sin\frac{\pi}{6} = \frac{\sqrt{3}}{2} + \frac{i}{2}$$

The first solution is $x_1 + iy_1 = \frac{\sqrt{3}}{2} + \frac{i}{2}$. The second and third solutions are:

$$x_2 = x_1 \cos\left(\frac{2\pi}{3}\right) - y_1 \sin\left(\frac{2\pi}{3}\right) = \frac{\sqrt{3}}{2}\left(-\frac{1}{2}\right) - \frac{1}{2}\left(\frac{\sqrt{3}}{2}\right) = -\frac{\sqrt{3}}{4} - \frac{\sqrt{3}}{4} = -\frac{\sqrt{3}}{2}$$

$$y_2 = x_1 \sin\left(\frac{2\pi}{3}\right) + y_1 \cos\left(\frac{2\pi}{3}\right) = \frac{\sqrt{3}}{2}\left(\frac{\sqrt{3}}{2}\right) + \frac{1}{2}\left(-\frac{1}{2}\right) = \frac{3}{4} - \frac{1}{4} = \frac{2}{4} = \frac{1}{2}$$

$$x_2 + iy_2 = -\frac{\sqrt{3}}{2} + \frac{i}{2}$$

$$x_3 = x_1 \cos\left(\frac{4\pi}{3}\right) - y_1 \sin\left(\frac{4\pi}{3}\right) = \frac{\sqrt{3}}{2}\left(-\frac{1}{2}\right) - \frac{1}{2}\left(-\frac{\sqrt{3}}{2}\right) = -\frac{\sqrt{3}}{4} + \frac{\sqrt{3}}{4} = 0$$

$$y_3 = x_1 \sin\left(\frac{4\pi}{3}\right) + y_1 \cos\left(\frac{4\pi}{3}\right) = \frac{\sqrt{3}}{2}\left(-\frac{\sqrt{3}}{2}\right) + \frac{1}{2}\left(-\frac{1}{2}\right) = -\frac{3}{4} - \frac{1}{4} = -1$$

$$x_3 + iy_3 = 0 - i = -i$$

Check: $\left(\frac{\sqrt{3}}{2} + \frac{i}{2}\right)\left(\frac{\sqrt{3}}{2} + \frac{i}{2}\right) = \frac{3}{4} + \frac{\sqrt{3}}{4}i + \frac{\sqrt{3}}{4}i + \frac{i^2}{4} = \frac{3}{4} + \frac{\sqrt{3}}{2}i - \frac{1}{4} = \frac{1}{2} + \frac{\sqrt{3}}{2}i$

$\left(\frac{1}{2} + \frac{\sqrt{3}}{2}i\right)\left(\frac{\sqrt{3}}{2} + \frac{i}{2}\right) = \frac{\sqrt{3}}{4} + \frac{i}{4} + \frac{3}{4}i + \frac{\sqrt{3}}{4}i^2 = \frac{\sqrt{3}}{4} + i - \frac{\sqrt{3}}{4} = i$

$\left(-\frac{\sqrt{3}}{2} + \frac{i}{2}\right)\left(-\frac{\sqrt{3}}{2} + \frac{i}{2}\right) = \frac{3}{4} - \frac{\sqrt{3}}{4}i - \frac{\sqrt{3}}{4}i + \frac{i^2}{4} = \frac{3}{4} - \frac{\sqrt{3}}{2}i - \frac{1}{4} = \frac{1}{2} - \frac{\sqrt{3}}{2}i$

$\left(\frac{1}{2} - \frac{\sqrt{3}}{2}i\right)\left(-\frac{\sqrt{3}}{2} + \frac{i}{2}\right) = -\frac{\sqrt{3}}{4} + \frac{i}{4} + \frac{3}{4}i - \frac{\sqrt{3}}{4}i^2 = -\frac{\sqrt{3}}{4} + i + \frac{\sqrt{3}}{4} = i$

$(-i)^3 = (-i)(-i)(-i) = (-1)^3 i^3 = -i^3 = -(-i) = i$

Notes: $(-1)^3 = -1$. Recall from Chapter 1 that $i^3 = -i$.

(7) Compare $-\sqrt{2} - i\sqrt{6}$ with $z = x + iy$ to see that $x = -\sqrt{2}$ and $y = -\sqrt{6}$.[2]

$$|z| = \sqrt{x^2 + y^2} = \sqrt{\left(-\sqrt{2}\right)^2 + \left(-\sqrt{6}\right)^2} = \sqrt{2 + 6} = \sqrt{8}$$

$$\theta = \tan^{-1}\left(\frac{y}{x}\right) = \tan^{-1}\left(\frac{-\sqrt{6}}{-\sqrt{2}}\right) = \tan^{-1}(\sqrt{3}) = \frac{\pi}{3} + \pi = \frac{4\pi}{3} \text{ rad (Quad. III)}$$

Notes: Add π rad to the calculator's answer for inverse tangent if $x < 0$ (Chapter 7).

---

[2] You should realize that $\sqrt{8} = \sqrt{(4)(2)} = \sqrt{4}\sqrt{2} = 2\sqrt{2}$, but since we will be taking the fourth root of this shortly, there isn't a benefit to doing this right now.

Complex Numbers Essentials Math Workbook with Answers

Make sure your calculator is in **radians** mode.

$$\sqrt{8}^{1/4} e^{(4i\pi/3)/4} = \left(8^{1/2}\right)^{1/4} e^{4i\pi/12} = 8^{1/8} e^{i\pi/3}$$

Notes: Since $|z| = \sqrt{8}$, it follows that $|z|^{1/4} = \sqrt{8}^{1/4} = \left(8^{1/2}\right)^{1/4} = 8^{1/8}$ according to the rule $(a^b)^c = a^{bc}$ since $\frac{1}{2}\left(\frac{1}{4}\right) = \frac{1}{8}$. It similarly follows that $e^{(4i\pi/3)/4} = e^{4i\pi/12} = e^{i\pi/3}$ since $\frac{4}{3}\left(\frac{1}{4}\right) = \frac{1}{3}$.

$$8^{1/8} e^{i\pi/3} = 8^{1/8} \cos\frac{\pi}{3} + i 8^{1/8} \sin\frac{\pi}{3} = 8^{1/8}\left(\frac{1}{2}\right) + i 8^{1/8}\left(\frac{\sqrt{3}}{2}\right)$$

$$= \frac{8^{1/8}}{2} + i\frac{8^{1/8}\sqrt{3}}{2} = \frac{2^{3/8}}{2} + i\frac{2^{3/8}\sqrt{3}}{2}$$

Alternate answer: $\frac{1}{2^{5/8}} + i\frac{\sqrt{3}}{2^{5/8}}$ (the above answer with a rational denominator is preferred). Note: $8^{1/8} = (2^3)^{1/8} = 2^{3/8}$ because $2^3 = 8$.

The first solution is $x_1 + iy_1 = \frac{2^{3/8}}{2} + i\frac{2^{3/8}\sqrt{3}}{2}$. The 2nd, 3rd, and 4th solutions are:

$$x_2 = x_1 \cos\left(\frac{2\pi}{4}\right) - y_1 \sin\left(\frac{2\pi}{4}\right) = \frac{2^{3/8}}{2}(0) - \frac{2^{3/8}\sqrt{3}}{2}(1) = -\frac{2^{3/8}\sqrt{3}}{2}$$

$$y_2 = x_1 \sin\left(\frac{2\pi}{4}\right) + y_1 \cos\left(\frac{2\pi}{4}\right) = \frac{2^{3/8}}{2}(1) + \frac{2^{3/8}\sqrt{3}}{2}(0) = \frac{2^{3/8}}{2}$$

$$x_2 + iy_2 = -\frac{2^{3/8}\sqrt{3}}{2} + \frac{2^{3/8}}{2}i$$

$$x_3 = x_1 \cos\left(\frac{4\pi}{4}\right) - y_1 \sin\left(\frac{4\pi}{4}\right) = \frac{2^{3/8}}{2}(-1) - \frac{2^{3/8}\sqrt{3}}{2}(0) = -\frac{2^{3/8}}{2}$$

$$y_3 = x_1 \sin\left(\frac{4\pi}{4}\right) + y_1 \cos\left(\frac{4\pi}{4}\right) = \frac{2^{3/8}}{2}(0) + \frac{2^{3/8}\sqrt{3}}{2}(-1) = -\frac{2^{3/8}\sqrt{3}}{2}$$

$$x_3 + iy_3 = -\frac{2^{3/8}}{2} - \frac{2^{1/8}\sqrt{3}}{2}i$$

$$x_4 = x_1 \cos\left(\frac{6\pi}{4}\right) - y_1 \sin\left(\frac{6\pi}{4}\right) = \frac{2^{3/8}}{2}(0) - \frac{2^{3/8}\sqrt{3}}{2}(-1) = \frac{2^{3/8}\sqrt{3}}{2}$$

$$y_4 = x_1 \sin\left(\frac{6\pi}{4}\right) + y_1 \cos\left(\frac{6\pi}{4}\right) = \frac{2^{3/8}}{2}(-1) + \frac{2^{3/8}\sqrt{3}}{2}(0) = -\frac{2^{3/8}}{2}$$

$$x_4 + iy_4 = \frac{2^{3/8}\sqrt{3}}{2} - \frac{2^{3/8}}{2}i$$

Answer Key

Check: $\left(\frac{2^{3/8}}{2} + i\frac{2^{3/8}\sqrt{3}}{2}\right)\left(\frac{2^{3/8}}{2} + i\frac{2^{3/8}\sqrt{3}}{2}\right) = \frac{2^{3/8}}{2}\frac{2^{3/8}}{2} + \frac{2^{3/8}}{2}i\frac{2^{3/8}\sqrt{3}}{2} + i\frac{2^{3/8}\sqrt{3}}{2}\frac{2^{3/8}}{2} +$
$i\frac{2^{3/8}\sqrt{3}}{2}i\frac{2^{3/8}\sqrt{3}}{2} = \frac{2^{6/8}}{4} + i\frac{2^{6/8}\sqrt{3}}{4} + i\frac{2^{6/8}\sqrt{3}}{4} + \frac{2^{6/8}(3)}{4}i^2 = \frac{2^{3/4}}{4} + i\frac{2^{3/4}\sqrt{3}}{2} - \frac{2^{3/4}(3)}{4}$
$= -\frac{2^{3/4}}{2} + i\frac{2^{3/4}\sqrt{3}}{2}$

$\left(\frac{2^{3/8}}{2} + i\frac{2^{3/8}\sqrt{3}}{2}\right)^4 = \left(\frac{2^{3/8}}{2} + i\frac{2^{3/8}\sqrt{3}}{2}\right)^2 \left(\frac{2^{3/8}}{2} + i\frac{2^{3/8}\sqrt{3}}{2}\right)^2$
$= \left(-\frac{2^{3/4}}{2} + i\frac{2^{3/4}\sqrt{3}}{2}\right)\left(-\frac{2^{3/4}}{2} + i\frac{2^{3/4}\sqrt{3}}{2}\right) = \frac{2^{6/4}}{4} - i\frac{2^{6/4}\sqrt{3}}{4} - i\frac{2^{6/4}\sqrt{3}}{4} + \frac{2^{6/4}(3)}{4}i^2$
$= \frac{2^{3/2}}{4} - i\frac{2^{3/2}\sqrt{3}}{2} - \frac{2^{3/2}(3)}{4} = -\frac{2^{3/2}}{2} - i\frac{2^{3/2}\sqrt{3}}{2} = -\frac{2\sqrt{2}}{2} - i\frac{2\sqrt{2}\sqrt{3}}{2} = -\sqrt{2} - i\sqrt{6}$

Note: $2^{3/2} = (2^{1/2})^3 = (\sqrt{2})^3 = \sqrt{2}\sqrt{2}\sqrt{2} = 2\sqrt{2}$ according to $(a^b)^c = a^{bc}$.

(8) Note that $81^{1/4} = \sqrt[4]{81}$. Since 81 is purely real, simply take the fourth root of 81 to find that one solution is $\sqrt[4]{81} = 3$ because $3^4 = 81$. (If you proceed to solve this problem as usual, you should get $y = 0$, $|z| = 81$, $\theta = 0$, and $81^{1/4}e^{0/4} = 81^{1/4}e^0 = 81^{1/4}\cos 0 + i81^{1/4}\sin 0 = 3 + 0i$.) The first solution is $x_1 + iy_1 = 3 + 0i = 3$. The 2nd, 3rd, and 4th solutions are:

$$x_2 = x_1 \cos\left(\frac{2\pi}{4}\right) - y_1 \sin\left(\frac{2\pi}{4}\right) = 3(0) - 0(1) = 0$$

$$y_2 = x_1 \sin\left(\frac{2\pi}{4}\right) + y_1 \cos\left(\frac{2\pi}{4}\right) = 3(1) + 0(0) = 3$$

$$x_2 + iy_2 = 0 + 3i = 3i$$

$$x_3 = x_1 \cos\left(\frac{4\pi}{4}\right) - y_1 \sin\left(\frac{4\pi}{4}\right) = 3(-1) - 0(0) = -3$$

$$y_3 = x_1 \sin\left(\frac{4\pi}{4}\right) + y_1 \cos\left(\frac{4\pi}{4}\right) = 3(0) + 0(-1) = 0$$

$$x_3 + iy_3 = -3 + 0i = -3$$

$$x_4 = x_1 \cos\left(\frac{6\pi}{4}\right) - y_1 \sin\left(\frac{6\pi}{4}\right) = 3(0) - 0(-1) = 0$$

$$y_4 = x_1 \sin\left(\frac{6\pi}{4}\right) + y_1 \cos\left(\frac{6\pi}{4}\right) = 3(-1) + 0(0) = -3$$

$$x_4 + iy_4 = 0 - 3i$$

The four solutions are 3, $3i$, -3, and $-3i$.

Check: $3^4 = 81$, $(-3)^4 = (-1)^4 3^4 = 81$, $(3i)^4 = 3^4 i^4 = 81(1) = 81$, and $(-3i)^4 = (-1)^4 3^4 i^4 = (1)(81)(1) = 81$ Note: Recall from Chapter 1 that $i^4 = 1$.

(9) Compare -64 with $z = x + iy$ to see that $x = -64$ and $y = 0$.[3]

$$|z| = \sqrt{x^2 + y^2} = \sqrt{(-64)^2 + 0^2} = \sqrt{4096 + 0} = \sqrt{4096} = 64$$

$$\theta = \tan^{-1}\left(\frac{y}{x}\right) = \tan^{-1}\left(\frac{0}{-64}\right) = \tan^{-1}(0) = \pi \text{ rad (on the neg. x-axis)}$$

Notes: Add π rad to the calculator's answer for inverse tangent if $x < 0$ (Chapter 7). Make sure your calculator is in **radians** mode.

$$64^{1/3} e^{i\pi/3} = 64^{1/3} \cos\frac{\pi}{3} + i 64^{1/3} \sin\frac{\pi}{3} = 4\left(\frac{1}{2}\right) + 4i\left(\frac{\sqrt{3}}{2}\right) = 2 + 2i\sqrt{3}$$

The first solution is $x_1 + iy_1 = 2 + 2i\sqrt{3}$. The second and third solutions are:

$$x_2 = x_1 \cos\left(\frac{2\pi}{3}\right) - y_1 \sin\left(\frac{2\pi}{3}\right) = 2\left(-\frac{1}{2}\right) - 2\sqrt{3}\left(\frac{\sqrt{3}}{2}\right) = -1 - 3 = -4$$

$$y_2 = x_1 \sin\left(\frac{2\pi}{3}\right) + y_1 \cos\left(\frac{2\pi}{3}\right) = 2\left(\frac{\sqrt{3}}{2}\right) + 2\sqrt{3}\left(-\frac{1}{2}\right) = \sqrt{3} - \sqrt{3} = 0$$

$$x_2 + iy_2 = -4 + 0i = -4$$

$$x_3 = x_1 \cos\left(\frac{4\pi}{3}\right) - y_1 \sin\left(\frac{4\pi}{3}\right) = 2\left(-\frac{1}{2}\right) - 2\sqrt{3}\left(-\frac{\sqrt{3}}{2}\right) = -1 + 3 = 2$$

$$y_3 = x_1 \sin\left(\frac{4\pi}{3}\right) + y_1 \cos\left(\frac{4\pi}{3}\right) = 2\left(-\frac{\sqrt{3}}{2}\right) + 2\sqrt{3}\left(-\frac{1}{2}\right) = -\sqrt{3} - \sqrt{3} = -2\sqrt{3}$$

$$x_3 + iy_3 = 2 - 2i\sqrt{3}$$

Check: $(2 + 2i\sqrt{3})(2 + 2i\sqrt{3}) = 4 + 4i\sqrt{3} + 4i\sqrt{3} + 4i^2(3) = 4 + 8i\sqrt{3} - 12$
$= -8 + 8i\sqrt{3}$
$(-8 + 8i\sqrt{3})(2 + 2i\sqrt{3}) = -16 - 16i\sqrt{3} + 16i\sqrt{3} + 16i^2(3) = -16 - 48 = -64$
$(-4)^3 = (-1)^3 4^3 = -64$
$(2 - 2i\sqrt{3})(2 - 2i\sqrt{3}) = 4 - 4i\sqrt{3} - 4i\sqrt{3} + 4i^2(3) = 4 - 8i\sqrt{3} - 12$
$= -8 - 8i\sqrt{3}$
$(-8 - 8i\sqrt{3})(2 - 2i\sqrt{3}) = -16 + 16i\sqrt{3} - 16i\sqrt{3} + 16i^2(3) = -16 - 48 = -64$

---

[3] Since -64 is purely real, you could find that one solution is $\sqrt[3]{-64} = -4$ because $(-4)^3 = (-1)^3 4^3 = -64$, and then proceed to find the remaining solutions in the usual way. In this case, you would call $x_1 + iy_1 = -4 + 0i$ and use this to find the other solutions.

Answer Key

(10) Compare $-208 - 144i$ with $z = x + iy$ to see that $x = -208$ and $y = -144$.[4]

$$|z| = \sqrt{x^2 + y^2} = \sqrt{(-208)^2 + (-144)^2} = \sqrt{43{,}264 + 20{,}736} = \sqrt{64{,}000}$$

$$\theta = \tan^{-1}\left(\frac{y}{x}\right) = \tan^{-1}\left(\frac{-144}{-208}\right) \approx 0.6055 + \pi \approx 3.7471 \text{ rad (Quad. III)}$$

Notes: Add π rad to the calculator's answer for inverse tangent if $x < 0$ (Chapter 7). Make sure your calculator is in **radians** mode.

$$\sqrt{64{,}000}^{2/3} e^{2(3.7471i)/3} \approx 64{,}000^{1/3} e^{2.498i} = 40 e^{2.498i}$$

Note: Since $|z| = \sqrt{64{,}000}$, it follows that $|z|^{2/3} = \sqrt{64{,}000}^{2/3} = \left(64{,}000^{1/2}\right)^{2/3} = 64{,}000^{1/3} = 40$ according to the rule $(a^b)^c = a^{bc}$ since $\frac{1}{2}\left(\frac{2}{3}\right) = \frac{2}{6} = \frac{1}{3}$.

$$40 e^{2.498i} = 40 \cos 2.498 + 40i \sin 2.498 \approx -32 + 24i$$

The first solution is $x_1 + iy_1 = -32 + 24i$. The second and third solutions are:

$$x_2 = x_1 \cos\left(\frac{2\pi}{3}\right) - y_1 \sin\left(\frac{2\pi}{3}\right) = -32\left(-\frac{1}{2}\right) - 24\left(\frac{\sqrt{3}}{2}\right) = 16 - 12\sqrt{3}$$

$$y_2 = x_1 \sin\left(\frac{2\pi}{3}\right) + y_1 \cos\left(\frac{2\pi}{3}\right) = -32\left(\frac{\sqrt{3}}{2}\right) + 24\left(-\frac{1}{2}\right) = -16\sqrt{3} - 12$$

$$x_2 + iy_2 = 16 - 12\sqrt{3} - 16i\sqrt{3} - 12i$$

$$x_3 = x_1 \cos\left(\frac{4\pi}{3}\right) - y_1 \sin\left(\frac{4\pi}{3}\right) = -32\left(-\frac{1}{2}\right) - 24\left(-\frac{\sqrt{3}}{2}\right) = 16 + 12\sqrt{3}$$

$$y_3 = x_1 \sin\left(\frac{4\pi}{3}\right) + y_1 \cos\left(\frac{4\pi}{3}\right) = -32\left(-\frac{\sqrt{3}}{2}\right) + 24\left(-\frac{1}{2}\right) = 16\sqrt{3} - 12$$

$$x_3 + iy_3 = 16 + 12\sqrt{3} + 16i\sqrt{3} - 12i$$

Check: For the first solution, we wish to verify that $(-208 - 144i)^{2/3} = -32 + 24i$. If we cube both sides, we get $(-208 - 144i)^2 = (-32 + 24i)^3$. So we just need to check if $(-208 - 144i)^2$ is equal to $(-32 + 24i)^3$.

$(-208 - 144i)(-208 - 144i) = 43{,}264 + 29{,}952i + 29{,}952i + 20{,}736i^2$
$= 43{,}264 + 59{,}904i - 20{,}736 = 22{,}528 + 59{,}904i$
$(-32 + 24i)(-32 + 24i) = 1024 - 768i - 768i + 576i^2$

---

[4] You could factor this as $\sqrt{64{,}000} = \sqrt{(6400)(10)} = 80\sqrt{10}$, but since we'll be raising this to the power of 2/3 soon, there isn't any benefit to doing this yet.

$= 1024 - 1536i - 576 = 448 - 1536i$

$(448 - 1536i)(-32 + 24i) = -14{,}336 + 10{,}752i + 49{,}152i - 36{,}864i^2$

$= -14{,}336 + 59{,}904i + 36{,}864 = 22{,}528 + 59{,}904i$ (smile!)

(If you wish to check the other solutions, it is simpler if you work with decimals on your calculator instead of carrying out algebra with the square roots.)

(11) Compare $-i$ with $z = x + iy$ to see that $x = 0$ and $y = -1$.

$$|z| = \sqrt{x^2 + y^2} = \sqrt{0^2 + (-1)^2} = \sqrt{0+1} = \sqrt{1} = 1$$

$$\theta = \tan^{-1}\left(\frac{y}{x}\right) = \tan^{-1}\left(\frac{-1}{0}\right) = \frac{3\pi}{2} \text{ or } -\frac{\pi}{2} \text{ rad (on the neg. } y\text{-axis)}$$

Since $x = 0$ and $y = -1$, the point lies on the $-y$-axis, which has an angle of $270°$ (equivalent to $\frac{3\pi}{2}$ rad). See the solution to Chapter 7, Exercise 9.

$$1^{5/6} e^{5(3i\pi/2)/6} = 1 e^{5i\pi/4} = e^{5i\pi/4}$$

Note: $e^{5(3i\pi/2)/6} = e^{5i\pi/4}$ according to the rule $(a^b)^c = a^{bc}$ since $\frac{5}{6}\left(\frac{3}{2}\right) = \frac{15}{12} = \frac{5}{4}$.

$$e^{5i\pi/4} = \cos\frac{5\pi}{4} + i\sin\frac{5\pi}{4} = -\frac{\sqrt{2}}{2} - \frac{\sqrt{2}}{2}i \text{ alternate answer: } -\frac{1}{\sqrt{2}} - \frac{i}{\sqrt{2}}$$

The first solution is $x_1 + iy_1 = -\frac{\sqrt{2}}{2} - \frac{\sqrt{2}}{2}i$. The 2nd thru 6th solutions are:

$$x_2 = x_1 \cos\left(\frac{2\pi}{6}\right) - y_1 \sin\left(\frac{2\pi}{6}\right) = -\frac{\sqrt{2}}{2}\left(\frac{1}{2}\right) - \left(-\frac{\sqrt{2}}{2}\right)\left(\frac{\sqrt{3}}{2}\right) = -\frac{\sqrt{2}}{4} + \frac{\sqrt{6}}{4}$$

$$y_2 = x_1 \sin\left(\frac{2\pi}{6}\right) + y_1 \cos\left(\frac{2\pi}{6}\right) = -\frac{\sqrt{2}}{2}\left(\frac{\sqrt{3}}{2}\right) + \left(-\frac{\sqrt{2}}{2}\right)\left(\frac{1}{2}\right) = -\frac{\sqrt{6}}{4} - \frac{\sqrt{2}}{4}$$

$$x_2 + iy_2 = -\frac{\sqrt{2}}{4} + \frac{\sqrt{6}}{4} - \frac{\sqrt{6}}{4}i - \frac{\sqrt{2}}{4}i$$

$$x_3 = x_1 \cos\left(\frac{4\pi}{6}\right) - y_1 \sin\left(\frac{4\pi}{6}\right) = -\frac{\sqrt{2}}{2}\left(-\frac{1}{2}\right) - \left(-\frac{\sqrt{2}}{2}\right)\left(\frac{\sqrt{3}}{2}\right) = \frac{\sqrt{2}}{4} + \frac{\sqrt{6}}{4}$$

$$y_3 = x_1 \sin\left(\frac{4\pi}{6}\right) + y_1 \cos\left(\frac{4\pi}{6}\right) = -\frac{\sqrt{2}}{2}\left(\frac{\sqrt{3}}{2}\right) + \left(-\frac{\sqrt{2}}{2}\right)\left(-\frac{1}{2}\right) = -\frac{\sqrt{6}}{4} + \frac{\sqrt{2}}{4}$$

$$x_3 + iy_3 = \frac{\sqrt{2}}{4} + \frac{\sqrt{6}}{4} - \frac{\sqrt{6}}{4}i + \frac{\sqrt{2}}{4}i$$

$$x_4 = x_1 \cos\left(\frac{6\pi}{6}\right) - y_1 \sin\left(\frac{6\pi}{6}\right) = -\frac{\sqrt{2}}{2}(-1) - \left(-\frac{\sqrt{2}}{2}\right)(0) = \frac{\sqrt{2}}{2} + 0 = \frac{\sqrt{2}}{2}$$

$$y_4 = x_1 \sin\left(\frac{6\pi}{6}\right) + y_1 \cos\left(\frac{6\pi}{6}\right) = -\frac{\sqrt{2}}{2}(0) + \left(-\frac{\sqrt{2}}{2}\right)(-1) = 0 + \frac{\sqrt{2}}{2} = \frac{\sqrt{2}}{2}$$

$$x_4 + iy_4 = \frac{\sqrt{2}}{2} + \frac{\sqrt{2}}{2}i$$

Answer Key

$x_5 = x_1 \cos\left(\frac{8\pi}{6}\right) - y_1 \sin\left(\frac{8\pi}{6}\right) = -\frac{\sqrt{2}}{2}\left(-\frac{1}{2}\right) - \left(-\frac{\sqrt{2}}{2}\right)\left(-\frac{\sqrt{3}}{2}\right) = \frac{\sqrt{2}}{4} - \frac{\sqrt{6}}{4}$

$y_5 = x_1 \sin\left(\frac{8\pi}{6}\right) + y_1 \cos\left(\frac{8\pi}{6}\right) = -\frac{\sqrt{2}}{2}\left(-\frac{\sqrt{3}}{2}\right) + \left(-\frac{\sqrt{2}}{2}\right)\left(-\frac{1}{2}\right) = \frac{\sqrt{6}}{4} + \frac{\sqrt{2}}{4}$

$x_5 + iy_5 = \frac{\sqrt{2}}{4} - \frac{\sqrt{6}}{4} + \frac{\sqrt{6}}{4}i + \frac{\sqrt{2}}{4}i$

$x_6 = x_1 \cos\left(\frac{10\pi}{6}\right) - y_1 \sin\left(\frac{10\pi}{6}\right) = -\frac{\sqrt{2}}{2}\left(\frac{1}{2}\right) - \left(-\frac{\sqrt{2}}{2}\right)\left(-\frac{\sqrt{3}}{2}\right) = -\frac{\sqrt{2}}{4} - \frac{\sqrt{6}}{4}$

$y_6 = x_1 \sin\left(\frac{10\pi}{6}\right) + y_1 \cos\left(\frac{10\pi}{6}\right) = -\frac{\sqrt{2}}{2}\left(-\frac{\sqrt{3}}{2}\right) + \left(-\frac{\sqrt{2}}{2}\right)\left(\frac{1}{2}\right) = \frac{\sqrt{6}}{4} - \frac{\sqrt{2}}{4}$

$x_6 + iy_6 = -\frac{\sqrt{2}}{4} - \frac{\sqrt{6}}{4} + \frac{\sqrt{6}}{4}i - \frac{\sqrt{2}}{4}i$

Check: For the first solution, we wish to verify that $(-i)^{5/6} = -\frac{\sqrt{2}}{2} - \frac{\sqrt{2}}{2}i$. If we raise both sides to the 6th power, we get $(-i)^5 = \left(-\frac{\sqrt{2}}{2} - \frac{\sqrt{2}}{2}i\right)^6$. So we just need to check if $(-i)^5$ is equal to $\left(-\frac{\sqrt{2}}{2} - \frac{\sqrt{2}}{2}i\right)^6$.

$(-i)^5 = (-1)^5 i^5 = (-1)i^1 = -i$ (recall Chapter 1)

$\left(-\frac{\sqrt{2}}{2} - \frac{\sqrt{2}}{2}i\right)\left(-\frac{\sqrt{2}}{2} - \frac{\sqrt{2}}{2}i\right) = \frac{2}{4} + \frac{2}{4}i + \frac{2}{4}i + \frac{2}{4}i^2 = \frac{2}{4} + \frac{4}{4}i - \frac{2}{4} = i$

$\left(-\frac{\sqrt{2}}{2} - \frac{\sqrt{2}}{2}i\right)^6 = \left(-\frac{\sqrt{2}}{2} - \frac{\sqrt{2}}{2}i\right)^2\left(-\frac{\sqrt{2}}{2} - \frac{\sqrt{2}}{2}i\right)^2\left(-\frac{\sqrt{2}}{2} - \frac{\sqrt{2}}{2}i\right)^2 = (i)(i)(i) = i^3 = -i$

Since $(-i)^5 = -i$ agrees with $\left(-\frac{\sqrt{2}}{2} - \frac{\sqrt{2}}{2}i\right)^6 = -i$, the first solution checks out.

(12) Compare $8 - 6i$ with $z = x + iy$ to see that $x = 8$ and $y = -6$.

$$|z| = \sqrt{x^2 + y^2} = \sqrt{8^2 + (-6)^2} = \sqrt{64 + 36} = \sqrt{100} = 10$$

$$\theta = \tan^{-1}\left(\frac{y}{x}\right) = \tan^{-1}\left(\frac{-6}{8}\right) \approx -0.6435 \text{ rad (Quad. IV)}$$

Make sure your calculator is in **radians** mode. (If you get -36.9, that's degrees.)

$$10^{5/2}e^{-5(0.6435i)/2} = 10^{5/2}e^{-1.6088i}$$

$10^{5/2}e^{-1.6088i} = 10^{5/2}\cos(-1.6088) + i10^{5/2}\sin(-1.6088) \approx -12 - 316i$

When the root involves a 1/2 (like this problem, where the power is 5/2), the second solution is the negative of the first solution: $12 + 316i$. (Only the denominator of the exponent determines the number of solutions. Here, there are 2 solutions because 5/2 has a denominator of 2.)

Check: For the second solution, we wish to verify that $(8 - 6i)^{5/2} = 12 + 316i$. If we square both sides, we get $(8 - 6i)^5 = (12 + 316i)^2$. So we just need to check if $(8 - 6i)^5$ is equal to $(12 + 316i)^2$.

$(8 - 6i)(8 - 6i) = 64 - 48i - 48i + 36i^2 = 64 - 96i - 36 = 28 - 96i$

$(8 - 6i)^4 = (8 - 6i)^2(8 - 6i)^2 = (28 - 96i)(28 - 96i)$

$= 784 - 2688i - 2688i + 9216i^2 = 784 - 5376i = -8432 - 5376i$

$(8 - 6i)^5 = (8 - 6i)^4(8 - 6i) = (-8432 - 5376i)(8 - 6i)$

$= -67{,}456 + 50{,}592i - 43{,}008 + 32{,}256i^2$

$= -67{,}456 + 7584i - 32{,}256 = -99{,}712 + 7584i$

$(12 + 316i)^2 = (12 + 316i)(12 + 316i) = 144 + 3792i + 3792i + 99{,}856i^2$

$= 144 + 3792i + 3792i + 99{,}856i^2 = 144 + 7584i - 99{,}712 = -99{,}712 + 7584i$

Chapter 14

(1) $z_1 = 1$, $z_2 = e^{2i\pi/2} = e^{i\pi} = \cos \pi + i \sin \pi = -1 + i(0) = -1$
Check: $1^2 = 1$ and $(-1)^2 = 1$

(2) $z_1 = 1$, $z_2 = e^{2i\pi/4} = e^{i\pi/2} = \cos\frac{\pi}{2} + i \sin\frac{\pi}{2} = 0 + i(1) = i$,

$z_3 = e^{4i\pi/4} = e^{i\pi} = \cos \pi + i \sin \pi = -1 + i(0) = -1$,

$z_4 = e^{6i\pi/4} = e^{3i\pi/2} = \cos\left(\frac{3\pi}{2}\right) + i \sin\left(\frac{3\pi}{2}\right) = 0 + i(-1) = -i$

Check: $1^4 = 1$, $i^4 = 1$, $(-1)^4 = 1$, and $(-i)^4 = (-1)^4 i^4 = (1)(1) = 1$

(3) $z_1 = 1$, $z_2 = e^{2i\pi/6} = e^{i\pi/3} = \cos\frac{\pi}{3} + i \sin\frac{\pi}{3} = \frac{1}{2} + \frac{\sqrt{3}}{2}i$,

$z_3 = e^{4i\pi/6} = e^{2i\pi/3} = \cos\left(\frac{2\pi}{3}\right) + i \sin\left(\frac{2\pi}{3}\right) = -\frac{1}{2} + \frac{\sqrt{3}}{2}i$,

$z_4 = e^{6i\pi/6} = e^{i\pi} = \cos \pi + i \sin \pi = -1 + i(0) = -1$,

$z_5 = e^{8i\pi/6} = e^{4i\pi/3} = \cos\left(\frac{4\pi}{3}\right) + i \sin\left(\frac{4\pi}{3}\right) = -\frac{1}{2} - \frac{\sqrt{3}}{2}i$,

$z_6 = e^{10i\pi/6} = e^{5i\pi/3} = \cos\left(\frac{2\pi}{3}\right) + i \sin\left(\frac{2\pi}{3}\right) = \frac{1}{2} - \frac{\sqrt{3}}{2}i$,

Check: $1^6 = 1$, $(-1)^6 = 1$,

$\left(\frac{1}{2} + \frac{\sqrt{3}}{2}i\right)^2 = \frac{1}{4} + \frac{\sqrt{3}}{4}i + \frac{\sqrt{3}}{4}i + \frac{3}{4}i^2 = \frac{1}{4} + \frac{\sqrt{3}}{2}i - \frac{3}{4} = -\frac{1}{2} + \frac{\sqrt{3}}{2}i$

$\left(\frac{1}{2} + \frac{\sqrt{3}}{2}i\right)^4 = \left(\frac{1}{2} + \frac{\sqrt{3}}{2}i\right)^2\left(\frac{1}{2} + \frac{\sqrt{3}}{2}i\right)^2 = \left(-\frac{1}{2} + \frac{\sqrt{3}}{2}i\right)\left(-\frac{1}{2} + \frac{\sqrt{3}}{2}i\right)$

$= \frac{1}{4} - \frac{\sqrt{3}}{4}i - \frac{\sqrt{3}}{4}i + \frac{3}{4}i^2 = \frac{1}{4} - \frac{\sqrt{3}}{2}i - \frac{3}{4} = -\frac{1}{2} - \frac{\sqrt{3}}{2}i$

$\left(\frac{1}{2} + \frac{\sqrt{3}}{2}i\right)^6 = \left(\frac{1}{2} + \frac{\sqrt{3}}{2}i\right)^4\left(\frac{1}{2} + \frac{\sqrt{3}}{2}i\right)^2 = \left(-\frac{1}{2} - \frac{\sqrt{3}}{2}i\right)\left(-\frac{1}{2} + \frac{\sqrt{3}}{2}i\right)$

$= \frac{1}{4} - \frac{\sqrt{3}}{4}i + \frac{\sqrt{3}}{4}i - \frac{3}{4}i^2 = \frac{1}{4} - \frac{3}{4}(-1) = \frac{1}{4} + \frac{3}{4} = 1$

(4) $z_1 = 1$, $z_2 = e^{2i\pi/12} = e^{i\pi/6} = \cos\left(\frac{\pi}{6}\right) + i \sin\frac{\pi}{6} = \frac{\sqrt{3}}{2} + \frac{1}{2}i$,

$z_3 = e^{4i\pi/12} = e^{i\pi/3} = \cos\frac{\pi}{3} + i \sin\frac{\pi}{3} = \frac{1}{2} + \frac{\sqrt{3}}{2}i$,

$z_4 = e^{6i\pi/12} = e^{i\pi/2} = \cos\frac{\pi}{2} + i \sin\frac{\pi}{2} = 0 + i(1) = i$,

$z_5 = e^{8i\pi/12} = e^{2i\pi/3} = \cos\left(\frac{2\pi}{3}\right) + i \sin\left(\frac{2\pi}{3}\right) = -\frac{1}{2} + \frac{\sqrt{3}}{2}i$,

$z_6 = e^{10i\pi/12} = e^{5i\pi/6} = \cos\left(\frac{5\pi}{6}\right) + i \sin\left(\frac{5\pi}{6}\right) = -\frac{\sqrt{3}}{2} + \frac{1}{2}i$,

$z_7 = e^{12i\pi/12} = e^{i\pi} = \cos \pi + i \sin \pi = -1 + i(0) = -1,$

$z_8 = e^{14i\pi/12} = e^{7i\pi/6} = \cos\left(\frac{7\pi}{6}\right) + i \sin\left(\frac{7\pi}{6}\right) = -\frac{\sqrt{3}}{2} - \frac{1}{2}i,$

$z_9 = e^{16i\pi/12} = e^{4i\pi/3} = \cos\left(\frac{4\pi}{3}\right) + i \sin\left(\frac{4\pi}{3}\right) = -\frac{1}{2} - \frac{\sqrt{3}}{2}i,$

$z_{10} = e^{18i\pi/12} = e^{3i\pi/2} = \cos\left(\frac{3\pi}{2}\right) + i \sin\left(\frac{3\pi}{2}\right) = 0 + i(-1) = -i,$

$z_{11} = e^{20i\pi/12} = e^{5i\pi/3} = \cos\left(\frac{5\pi}{3}\right) + i \sin\left(\frac{5\pi}{3}\right) = \frac{1}{2} - \frac{\sqrt{3}}{2}i,$

$z_{12} = e^{22i\pi/12} = e^{11i\pi/6} = \cos\left(\frac{11\pi}{6}\right) + i \sin\left(\frac{11\pi}{6}\right) = \frac{\sqrt{3}}{2} - \frac{1}{2}i,$

Check: $1^{12} = 1, i^{12} = (i^4)^3 = 1^3 = 1, (-1)^{12} = 1, (-i)^{12} = (-1)^{12} i^{12} = (1)(1) = 1,$

$\left(\frac{\sqrt{3}}{2} + \frac{1}{2}i\right)^2 = \frac{3}{4} + \frac{\sqrt{3}}{4}i + \frac{\sqrt{3}}{4}i + \frac{1}{4}i^2 = \frac{3}{4} + \frac{\sqrt{3}}{2}i - \frac{1}{4} = \frac{1}{2} + \frac{\sqrt{3}}{2}i$

$\left(\frac{\sqrt{3}}{2} + \frac{1}{2}i\right)^4 = \left(\frac{\sqrt{3}}{2} + \frac{1}{2}i\right)^2 \left(\frac{\sqrt{3}}{2} + \frac{1}{2}i\right)^2 = \left(\frac{1}{2} + \frac{\sqrt{3}}{2}i\right)\left(\frac{1}{2} + \frac{\sqrt{3}}{2}i\right)$

$= \frac{1}{4} + \frac{\sqrt{3}}{4}i + \frac{\sqrt{3}}{4}i + \frac{3}{4}i^2 = \frac{1}{4} + \frac{\sqrt{3}}{2}i - \frac{3}{4} = -\frac{1}{2} + \frac{\sqrt{3}}{2}i$

$\left(\frac{\sqrt{3}}{2} + \frac{1}{2}i\right)^6 = \left(\frac{\sqrt{3}}{2} + \frac{1}{2}i\right)^4 \left(\frac{\sqrt{3}}{2} + \frac{1}{2}i\right)^2 = \left(-\frac{1}{2} + \frac{\sqrt{3}}{2}i\right)\left(\frac{1}{2} + \frac{\sqrt{3}}{2}i\right)$

$= -\frac{1}{4} - \frac{\sqrt{3}}{4}i + \frac{\sqrt{3}}{4}i + \frac{3}{4}i^2 = -\frac{1}{4} - \frac{3}{4} = -\frac{1}{4} - \frac{3}{4} = -1$

$\left(\frac{\sqrt{3}}{2} + \frac{1}{2}i\right)^{12} = \left(\frac{\sqrt{3}}{2} + \frac{1}{2}i\right)^6 \left(\frac{\sqrt{3}}{2} + \frac{1}{2}i\right)^6 = (-1)(-1) = 1$

$\left(\frac{1}{2} + \frac{\sqrt{3}}{2}i\right)^2 = \frac{1}{4} + \frac{\sqrt{3}}{4}i + \frac{\sqrt{3}}{4}i + \frac{3}{4}i^2 = \frac{1}{4} + \frac{\sqrt{3}}{2}i - \frac{3}{4} = -\frac{1}{2} + \frac{\sqrt{3}}{2}i$

$\left(\frac{1}{2} + \frac{\sqrt{3}}{2}i\right)^4 = \left(\frac{1}{2} + \frac{\sqrt{3}}{2}i\right)^2 \left(\frac{1}{2} + \frac{\sqrt{3}}{2}i\right)^2 = \left(-\frac{1}{2} + \frac{\sqrt{3}}{2}i\right)\left(-\frac{1}{2} + \frac{\sqrt{3}}{2}i\right)$

$= \frac{1}{4} - \frac{\sqrt{3}}{4}i - \frac{\sqrt{3}}{4}i + \frac{3}{4}i^2 = \frac{1}{4} - \frac{\sqrt{3}}{2}i - \frac{3}{4} = -\frac{1}{2} - \frac{\sqrt{3}}{2}i$

$\left(\frac{1}{2} + \frac{\sqrt{3}}{2}i\right)^6 = \left(\frac{1}{2} + \frac{\sqrt{3}}{2}i\right)^4 \left(\frac{1}{2} + \frac{\sqrt{3}}{2}i\right)^2 = \left(-\frac{1}{2} - \frac{\sqrt{3}}{2}i\right)\left(-\frac{1}{2} + \frac{\sqrt{3}}{2}i\right)$

$= \frac{1}{4} - \frac{\sqrt{3}}{4}i + \frac{\sqrt{3}}{4}i - \frac{3}{4}i^2 = \frac{1}{4} - \frac{3}{4}(-1) = \frac{1}{4} + \frac{3}{4} = 1$

$\left(\frac{\sqrt{3}}{2} + \frac{1}{2}i\right)^{12} = \left(\frac{\sqrt{3}}{2} + \frac{1}{2}i\right)^6 \left(\frac{\sqrt{3}}{2} + \frac{1}{2}i\right)^6 = (1)(1) = 1$

Answer Key

Chapter 15

(1) $z = 7$

Check: $p(7) = 6(7) - 42 = 42 - 42 = 0$

(2) $z = 4, z = 5$, or $z = 9$

Notes: $a_0 = 180$ so the possible integer roots are $\pm 1, 2, 3, 4, 5, 6, 9, 10, 12, 15, 18, 20, 30, 36, 45, 60, 90$, and 180. Fortunately, you can find all 3 roots by the time you reach 9.

Note about checking the answers: You can plug each value of z into $p(z)$ and verify that $p(z)$ equals zero for each root like we did in the examples. An alternative way to check the answers is to multiply out the factored form and verify that it reproduces the given polynomial. We'll use the alternate method in this answer key.

Check: $(z-4)(z-5)(z-9) = (z^2 - 9z + 20)(z-9)$
$= z^3 - 9z^2 + 20z - 9z^2 + 81z - 180 = z^3 - 18z^2 + 101z - 180 = p(z)$

(3) $z = -\dfrac{2}{3}, z = 0, z = \dfrac{1}{4}$, or $z = 6$

Notes: Since $a_0 = 0$ (there isn't any constant term), one solution is $z = 0$.
Factor out z to get the simpler polynomial $12z^3 - 67z^2 - 32z + 12$.
In this simpler polynomial, the constant term is 12 so the possible integer roots are $\pm 1, 2, 3, 4, 6$, and 12. Of these, only 6 works.
In the simpler polynomial, the leading coefficient and constant term are both 12, so the possible numerators and denominators are 1, 2, 3, 4, 6, and 12. (As in the examples, discard 1 for the denominator.) The possible **reduced** fractions are
$\pm \dfrac{1}{12}, \dfrac{1}{6}, \dfrac{1}{4}, \dfrac{1}{3}, \dfrac{1}{2}, \dfrac{2}{3}$, and $\dfrac{3}{4}$. (Note that $\dfrac{2}{4} = \dfrac{1}{2}, \dfrac{2}{6} = \dfrac{1}{3}, \dfrac{2}{12} = \dfrac{1}{6}, \dfrac{3}{6} = \dfrac{1}{2}, \dfrac{3}{12} = \dfrac{1}{4}, \dfrac{4}{6} = \dfrac{2}{3}, \dfrac{4}{12} = \dfrac{1}{3}$, and $\dfrac{6}{12} = \dfrac{1}{2}$, which have all already been listed.) Of these, $\dfrac{1}{4}$ and $-\dfrac{2}{3}$ work.

Check: $(3z+2)(4z-1)z(z-6) = (12z^2 + 5z - 2)(z^2 - 6z)$
$= 12z^4 + 5z^3 - 2z^2 - 72z^3 - 30z^2 + 12z = 12z^4 - 67z^3 - 32z^2 + 12z = p(z)$

(4) $z = 1$ or $z = \dfrac{5}{4}$ (this one is a double root)

Notes: $a_0 = -25$ so the possible integer roots are $\pm 1, 5$, and 25. Of these, only 1 works. $a_3 = 16$ so the possible denominators are 4 and 16; $a_0 = -25$ so the possible numerators are 1, 5, and 25. The possible **reduced** fractions are $\pm \dfrac{1}{16}, \dfrac{1}{4}, \dfrac{5}{16}, \dfrac{5}{4}, \dfrac{25}{16}, \dfrac{25}{4}$.

Of these, only $\dfrac{5}{4}$ works.

Since the coefficients are real, if any roots are irrational or complex, they would come as part of a conjugate pair. But since we already found 2 out of the 3 roots, there can't be a pair of roots remaining, which tells us that one of our answers is a double root. To see which, factor $(z-1)$ out of $p(z)$. If you use the method of Footnote 4, you should get $(z-1)(az^2 + bz + c) = az^3 + (b-a)z^2 + (c-b)z - c$. Compare this with $16z^3 - 56z^2 + 65z - 25$ to identify $a = 16$ and $c = 25$. Either of the other terms will lead to $b = -40$. Therefore, the factorization is $(z-1)(16z^2 - 40z + 25)$. The quadratic factors as $(4z-5)(4z-5)$, which shows that $\frac{5}{4}$ is the double root.

Check: $(z-1)(4z-5)(4z-5) = (z-1)(16z^2 - 40z + 25)$
$= 16z^3 - 40z^2 + 25z - 16z^2 + 40z - 25 = 16z^3 - 56z^2 + 65z - 25 = p(z)$

(5) $z = -3, z = -2, z = \frac{5}{6}, z = 2$, or $z = 3$

Notes: $a_0 = -36$ so the possible integer roots are $\pm 1, 2, 3, 4, 6, 9, 12, 18,$ or 36. Of these, ± 2 and ± 3 work.

$a_5 = 6$ so the possible denominators are 2, 3, and 6; $a_0 = -36$ so the possible numerators are 1, 2, 3, 4, 6, 9, 12, 18, or 36. Of the many possible reduced fractions, only $\frac{5}{6}$ works. That's a lot of choices, but having already found 4 integer roots, we know that there is only one root left. If you can factor out $(z-2), (z+2), (z+3)$, and $(z-3)$, then you'll be left with $(6z-5)$; that is, factoring will give you the answer without having to check several different fractions.

Check: $(z-2)(z+2)(z-3)(z+3)(6z-5) = (z^2-4)(z^2-9)(6z-5) =$
$(z^4 - 13z^2 + 36)(6z-5) = 6z^5 - 5z^4 - 78z^3 + 65z^2 + 216z - 180 = p(z)$

(6) $z = -1, z = -\sqrt{3}$, or $z = \sqrt{3}$ (note that $\pm\sqrt{3}$ form a conjugate pair)

Notes: Since the coefficients are 1, 1, -3, and -3, this suggests that -1 or 1 might be roots. If you check -1, you'll find that it is indeed a root.

Factor out $(z+1)$ to reduce the cubic to a quadratic. If you use the method of Footnote 4, you should get $(z+1)(az^2 + bz + c) = az^3 + (a+b)z^2 + (b+c)z + c$. Identify $a = 1$ and $c = -3$. Either of the other terms will lead to $b = 0$. Therefore, the factorization is $(z+1)(z^2 - 3)$. Set $z^2 - 3$ equal to 0 to find that $z = \pm\sqrt{3}$.

Check: $(z+1)(z-\sqrt{3})(z+\sqrt{3}) = (z+1)(z^2 - 3) = z^3 + z^2 - 3z - 3 = p(z)$

Answer Key

(7) $z = -\frac{3}{2}$, $z = -2i$, or $z = 2i$ (note that $\pm 2i$ are complex conjugates)

Notes: $a_0 = 12$ so the possible integer roots are $\pm 1, 2, 3, 4, 6,$ and 12. None work. $a_3 = 2$ so the denominator must be 2; $a_0 = 12$ so the possible numerators are 1, 2, 3, 4, 6, and 12. The possible **reduced** fractions are $\pm\frac{1}{2}$ or $\pm\frac{3}{2}$ (because other combinations like $\frac{6}{2}$ reduce to integers). Of these, only $-\frac{3}{2}$ works.

Since there is a single integer or fractional root, the remaining two roots must be a conjugate pair (either complex or irrational conjugates). To find them, factor out $(2z + 3)$. If you use the method of Footnote 4, you should get $(2z + 3)(az^2 + bz + c) = 2az^3 + (3a + 2b)z^2 + (3b + 2c)z + 3c$. Identify $a = 1$ and $c = 4$. Either of the other terms will lead to $b = 0$. Therefore, the factorization is $(2z + 3)(z^2 + 4)$. Set $z^2 + 4$ equal to zero to find that $z = \pm\sqrt{-4} = \pm 2i$.

Check: $(2z + 3)(z - 2i)(z + 2i) = (2z + 3)(z^2 + 4) = 2z^3 + 3z^2 + 8z + 12 = p(z)$

(8) $z = -4$, $z = 4$, $z = -\sqrt{3}$, $z = \sqrt{3}$, $z = -i$, or $z = i$ (there are 2 conjugate pairs)

Notes: $a_0 = 48$ so the possible integer roots are $\pm 1, 2, 3, 4, 6, 8, 12, 16, 24,$ and 48. Of these, only ± 4 work.

$a_6 = 1$ so any fraction would have a denominator of 1, which wouldn't be a fraction; it would be an integer. So there isn't a real fraction that is a root.

But there are still 4 more roots to find. If they aren't integers or fractions, they must be two conjugate pairs. How can you find all four?

The 'trick' is to look at the coefficients: 1, -18, 29, and 48. Note that 1, 18, and 29 add up to 48. This suggests that the terms might cancel easily somehow. We would first try ± 1, but neither of these work. Hey, note that all of the exponents are **even**. Recall from Chapter 1 that even powers of i are -1 or 1. This suggests trying $z = \pm i$. Both work:

$i^6 - 18i^4 + 29i^2 + 48 = i^4 i^2 - 18(1) + 29(-1) + 48 = (1)(-1) - 18 - 29 + 48 = -1 - 47 + 48 = 0$ (and similarly for $-i$).

To get the last two roots, first factor out $(z - 4)$ and $(z + 4)$. We can factor out both at once. Note that $(z - 4)(z + 4) = z^2 - 16$, so we just need to factor out $z^2 - 16$. If you use the method of Footnote 4, you should get $(z^2 - 16)(az^4 + bz^2 + c)$ $= az^6 + (b - 16a)z^4 + (c - 16b)z^2 - 16c$. (We know this polynomial will have degree 4 because you need to multiply a polynomial of degree 2 by a polynomial of degree 4 in order to make a polynomial of degree 6. Multiply a few polynomials of various degrees to realize this point. Also, since the product makes a polynomial with only even terms, we only need to multiply $z^2 - 16$ by even terms.) Identify $a = 1$ and $c = -3$. Either of the other terms will lead to $b = -2$. Therefore, the factorization is $(z^2 - 16)(z^4 - 2z^2 - 3)$. Now factor out $(z - i)(z + i) = z^2 - i^2 = z^2 - (-1) = z^2 + 1$. Again, we can factor out $z^2 + 1$ in a single step. This factorization is somewhat simpler: $(z^2 + 1)(z^2 - 3) = z^4 - 2z^2 - 3$. Set $z^2 - 3$ equal to zero to find that the remaining roots are $z = \pm\sqrt{3}$.

Check: $(z - 4)(z + 4)(z - \sqrt{3})(z + \sqrt{3})(z - i)(z + i) = (z^2 - 16)(z^2 - 3)(z^2 + 1) =$
$(z^4 - 19z^2 + 48)(z^2 + 1) = z^6 - 19z^4 + 48z^2 + z^4 - 19z^2 + 48$
$= z^6 - 18z^4 + 29z^2 + 48 = p(z)$

(9) $z = \frac{1}{6}$, $z = 4 - 3i$, or $z = 4 + 3i$ (note that $4 \pm 3i$ are complex conjugates)

Notes: $a_0 = -25$ so the possible integer roots are $\pm 1, 5$, and 25. None work. $a_3 = 6$ so the possible denominators are 2, 3, and 6; $a_0 = -25$ so the possible numerators are 1, 5, and 25. The possible **reduced** fractions are $\pm\frac{1}{6}, \frac{1}{3}, \frac{1}{2}, \frac{5}{6}, \frac{5}{3}, \frac{5}{2}, \frac{25}{6}, \frac{25}{3}$, or $\frac{25}{2}$ Of these, only $\frac{1}{6}$ works.

To find the remaining conjugate pair, factor out $(6z - 1)$. If you use the method of Footnote 4, you should get $(6z - 1)(az^2 + bz + c) =$
$6az^3 + (6b - a)z^2 + (6c - b)z - c$. Identify $a = 1$ and $c = 25$. Either of the other terms will lead to $b = -8$. Therefore, the factorization is $(6z - 1)(z^2 - 8z + 25)$. Set $z^2 - 8z + 25$ equal to zero to find the remaining roots. Use the quadratic formula.

$z = \frac{-(-8)\pm\sqrt{(-8)^2-4(1)(25)}}{2(1)} = \frac{8\pm\sqrt{64-100}}{2} = \frac{8\pm\sqrt{-36}}{2} = \frac{8\pm 6i}{2} = 4 \pm 3i$

Check: $(6z - 1)(z - 4 - 3i)(z - 4 + 3i) = (6z - 1)(z^2 - 8z + 25)$
$= 6z^3 - 48z^2 + 150z - z^2 + 8z - 25 = 6z^3 - 49z^2 + 158z - 25 = p(z)$

(10) $z = -2$, $z = -\frac{3}{4}$, $z = \frac{1}{4}$, or $z = 2i$ (but **not** $-2i$ this time; see the note below)

Notes: $a_0 = 12i$; even though it is imaginary, the possible integer roots are still ± 1, 2, 3, 4, 6, and 12. Of these, -2 works. (If you plug -2 into the given polynomial and carry out the math correctly, you'll see that both the real and imaginary parts add up to zero.)

Answer Key

Although there is an imaginary coefficient, the same technique for fractions still works. $a_4 = 16$ so the possible denominators are 2, 4, 8, and 16; $a_0 = 12i$ so the possible numerators are 1, 2, 3, 4, 6, and 12. The possible **reduced** fractions are $\pm \frac{1}{16}, \frac{1}{8}, \frac{1}{4}, \frac{1}{2}, \frac{3}{2}, \frac{3}{4}, \frac{3}{8},$ or $\frac{3}{16}$. (If the numerator is 2, 4, 6, or 12, we either get an integer or one of these same reduced fractions.) Of these, only $\frac{1}{4}$ and $-\frac{4}{3}$ work.

There is only one root left. The presence of the imaginary number in the coefficients suggests that the last root is complex or purely imaginary. We **won't** get a pair of complex conjugates here; that rule only applies when all coefficients are real. You could get the last root by factoring out $4z - 1$, $4z + 3$, and $z + 2$. Alternatively, looking at $a_0 = 12i$ suggests trying $\pm i, 2i, 3i, 4i, 6i,$ and $12i$. It turns out that the remaining root is $2i$. If your calculator doesn't handle complex arithmetic, you need to work these out by hand, but since the second possibility works, it's not too bad. Again, you could just find the complex root by factoring like we did for previous exercises.

Check: $(4z - 1)(4z + 3)(z - 2i)(z + 2) = (16z^2 + 8z - 3)(z^2 + 2z - 2iz - 4i)$
$= 16z^4 + 8(5 - 4i)z^3 + (13 - 80i)z^2 - 2(3 + 13i)z + 12i = p(z)$

Chapter 16

(1) $i^i = e^{-\pi/2 - 2\pi k}$, which includes to $e^{-\pi/2}, e^{-5\pi/2}, e^{-9\pi/2}$, etc. and $e^{3\pi/2}, e^{7\pi/2}$, etc.
Alternate answer: If you let $k' = -k$, you can express the answer as $e^{-\pi/2 + 2\pi k'}$.
Notes: Use $z^w = e^{w \ln z} = e^{w(i\theta + 2\pi i k + \ln|z|)}$. Compare z^w to i^i to see that $z = i$ and $w = i$.
For $z = i$, $|z| = 1$ and $\theta = \pi/2$. This agrees with $e^{i\pi/2} = \cos\frac{\pi}{2} + i\sin\frac{\pi}{2} = 0 + 1i = i$.
$i^i = e^{i(i\pi/2 + 2\pi i k + \ln 1)} = e^{i(i\pi/2 + 2\pi i k + 0)} = e^{i^2\pi/2 + 2\pi i^2 k} = e^{-\pi/2 - 2\pi k}$

(2) $(-1)^i = e^{-\pi - 2\pi k}$, which includes $e^{-\pi}, e^{-3\pi}, e^{-5\pi}$, etc. and $e^{\pi}, e^{3\pi}, e^{5\pi}$, etc.
Alternate answer: If you let $k' = -k$, you can express the answer as $e^{-\pi + 2\pi k'}$. You can even write $e^{\pi + 2\pi k''}$ if you let $k'' = k' - 2\pi$. Any of these k's is the full set of integers.
Notes: Use $z^w = e^{w \ln z} = e^{w(i\theta + 2\pi i k + \ln|z|)}$. Compare z^w to $(-1)^i$ to see that $z = -1$ and $w = i$. For $z = -1$, $|z| = 1$ and $\theta = \pi$. This agrees with $e^{i\pi} = \cos\pi + i\sin\pi = -1 + 0i = -1$.
$(-1)^i = e^{i(i\pi + 2\pi i k + \ln 1)} = e^{i(i\pi + 2\pi i k + 0)} = e^{i^2\pi + 2\pi i^2 k} = e^{-\pi - 2\pi k}$

(3) $\ln(i) = i\frac{\pi}{2} + 2\pi i k$, which includes $i\frac{\pi}{2}, \frac{5i\pi}{2}, \frac{9i\pi}{2}$, etc. and $-\frac{3i\pi}{2}, -\frac{7i\pi}{2}$, etc.
Notes: Use $\ln z = \ln(|z|e^{i\theta}) = i\theta + 2\pi i k + \ln|z|$. For $z = i$, $|z| = 1$ and $\theta = \frac{\pi}{2}$. This agrees with $e^{i\pi/2} = \cos\frac{\pi}{2} + i\sin\frac{\pi}{2} = 0 + 1i = i$.
$\ln i = i\frac{\pi}{2} + 2\pi i k + \ln|1| = i\frac{\pi}{2} + 2\pi i k + 0 = i\frac{\pi}{2} + 2\pi i k$

(4) $\ln(e^{i\pi}) = i\pi + 2\pi i k$, which includes $i\pi, 3i\pi, 5i\pi$, etc. and $-i\pi, -3i\pi, -5i\pi$, etc.
Notes: Use $\ln z = \ln(|z|e^{i\theta}) = i\theta + 2\pi i k + \ln|z|$. For $z = e^{i\pi}$ (which is exponential form), $|z| = 1$ and $\theta = \pi$.
$\ln(e^{i\pi}) = i\pi + 2\pi i k + \ln|1| = i\pi + 2\pi i k + 0 = i\pi + 2\pi i k$

(5) $\left(\frac{i}{2} - \frac{\sqrt{3}}{2}\right)^i = e^{-5\pi/6 - 2\pi k}$, which includes to $e^{-5\pi/6}, e^{-17\pi/6}, e^{-29\pi/6}$, etc. and $e^{7\pi/6}, e^{19\pi/6}, e^{31\pi/6}$, etc.
Alternate answer: If you let $k' = -k$, you can express the answer as $e^{-5\pi/6 + 2\pi k'}$.
Notes: Let $z = x + iy = -\frac{\sqrt{3}}{2} + \frac{i}{2}$. Then $x = -\frac{\sqrt{3}}{2}$ and $y = \frac{1}{2}$.
$|z| = \sqrt{x^2 + y^2} = \sqrt{\left(-\frac{\sqrt{3}}{2}\right)^2 + \left(\frac{1}{2}\right)^2} = \sqrt{\frac{3}{4} + \frac{1}{4}} = \sqrt{1} = 1$

Answer Key

$\theta = \tan^{-1}\left(\frac{y}{x}\right) = \tan^{-1}\left(\frac{1}{2} \div -\frac{\sqrt{3}}{2}\right) = \tan^{-1}\left(-\frac{1}{2} \div \frac{2}{\sqrt{3}}\right) = \tan^{-1}\left(-\frac{1}{\sqrt{3}}\right) = -\frac{\pi}{6} + \pi = \frac{5\pi}{6}$

Like we did in Chapter 7, add π radians to the calculator's answer when $x < 0$ in order to put the argument in the correct quadrant (which in this case is Quad. II).

Use $z^w = e^{w \ln z} = e^{w(i\theta + 2\pi i k + \ln|z|)}$. Compare z^w to $\left(\frac{i}{2} - \frac{\sqrt{3}}{2}\right)^i = z^i$ to see that $w = i$.

$\left(\frac{i}{2} - \frac{\sqrt{3}}{2}\right)^i = e^{i(i5\pi/6 + 2\pi i k + \ln 1)} = e^{i(i5\pi/6 + 2\pi i k + 0)} = e^{5i^2\pi/6 + 2\pi i^2 k} = e^{-5\pi/6 - 2\pi k}$

(6) $\ln\left(\frac{\sqrt{3}}{2} + \frac{i}{2}\right) = i\frac{\pi}{6} + 2\pi i k$, which includes $i\frac{\pi}{6}, \frac{13i\pi}{6}, \frac{25i\pi}{6}$, etc. and $-\frac{11i\pi}{6}, -\frac{23i\pi}{6}$, etc.

Notes: Let $z = x + iy = \frac{\sqrt{3}}{2} + \frac{i}{2}$. Then $x = \frac{\sqrt{3}}{2}$ and $y = \frac{1}{2}$.

$|z| = \sqrt{x^2 + y^2} = \sqrt{\left(\frac{\sqrt{3}}{2}\right)^2 + \left(\frac{1}{2}\right)^2} = \sqrt{\frac{3}{4} + \frac{1}{4}} = \sqrt{1} = 1$

$\theta = \tan^{-1}\left(\frac{y}{x}\right) = \tan^{-1}\left(\frac{1}{2} \div \frac{\sqrt{3}}{2}\right) = \tan^{-1}\left(\frac{1}{2} \div \frac{2}{\sqrt{3}}\right) = \tan^{-1}\left(\frac{1}{\sqrt{3}}\right) = \frac{\pi}{6}$

Notes: Use $\ln z = \ln(|z|e^{i\theta}) = i\theta + 2\pi i k + \ln|z|$.

$\ln\left(\frac{\sqrt{3}}{2} + \frac{i}{2}\right) = i\frac{\pi}{6} + 2\pi i k + \ln|1| = i\frac{\pi}{6} + 2\pi i k + 0 = i\frac{\pi}{6} + 2\pi i k$

(7) $\ln\left(\frac{\sqrt{2}}{2} - i\frac{\sqrt{2}}{2}\right) = -i\frac{\pi}{4} + 2\pi i k$, which includes $-\frac{i\pi}{4}, -\frac{9i\pi}{4}, -\frac{17i\pi}{4}$, etc. and $\frac{7i\pi}{4}, \frac{15i\pi}{4}$, etc.

Notes: Let $z = x + iy = \frac{\sqrt{2}}{2} - i\frac{\sqrt{2}}{2}$. Then $x = \frac{\sqrt{2}}{2}$ and $y = -\frac{\sqrt{2}}{2}$.

$|z| = \sqrt{x^2 + y^2} = \sqrt{\left(\frac{\sqrt{2}}{2}\right)^2 + \left(-\frac{\sqrt{2}}{2}\right)^2} = \sqrt{\frac{2}{4} + \frac{2}{4}} = \sqrt{1} = 1$

$\theta = \tan^{-1}\left(\frac{y}{x}\right) = \tan^{-1}\left(-\frac{\sqrt{2}}{2} \div \frac{\sqrt{2}}{2}\right) = \tan^{-1}\left(-\frac{\sqrt{2}}{2} \div \frac{2}{\sqrt{2}}\right) = \tan^{-1}(-1) = -\frac{\pi}{4}$

Notes: Use $\ln z = \ln(|z|e^{i\theta}) = i\theta + 2\pi i k + \ln|z|$.

$\ln\left(\frac{\sqrt{2}}{2} - i\frac{\sqrt{2}}{2}\right) = -i\frac{\pi}{4} + 2\pi i k + \ln|1| = -i\frac{\pi}{4} + 2\pi i k + 0 = -i\frac{\pi}{4} + 2\pi i k$

(8) $i^{3-2i} = -ie^{\pi + 4\pi k}$, which includes to $-ie^{\pi}, -ie^{5\pi}, -ie^{9\pi}$, etc. and $-ie^{-3\pi}, -ie^{-7\pi}$, etc. This answer is equivalent to $e^{3i\pi/2 + 6\pi i k + \pi + 4\pi k}$ (see below).

Notes: Use $z^w = e^{w \ln z} = e^{w(i\theta + 2\pi i k + \ln|z|)}$. Compare z^w to i^{3-2i} to see that $z = i$ and $w = 3 - 2i$.

For $z = i$, $|z| = 1$ and $\theta = \pi/2$. This agrees with $e^{i\pi/2} = \cos\frac{\pi}{2} + i\sin\frac{\pi}{2} = 0 + 1i = i$.

$i^{3-2i} = e^{(3-2i)(i\pi/2+2\pi ik+\ln 1)} = e^{(3-2i)(i\pi/2+2\pi ik+0)} = e^{(3-2i)(i\pi/2+2\pi ik)} =$

$e^{3i\pi/2+6\pi ik-i^2\pi-4\pi i^2 k} = e^{3i\pi/2+6\pi ik+\pi+4\pi k} = e^{3i\pi/2+6\pi ik}e^{\pi+4\pi k} = -ie^{\pi+4\pi k}$

Using the rule $e^a e^b = e^{a+b}$, one must be careful. For example, $e^a = e^{a+2i\pi k}$. If we allow for the $+2i\pi k$ in the exponent of $e^{3i\pi/2+6\pi ik}$, we still get $e^{3i\pi/2+6\pi ik} = -i$ since $e^{3i\pi/2} = \cos\left(\frac{3\pi}{2}\right) + i\sin\left(\frac{3\pi}{2}\right) = -i$ even if you add $2\pi k$ to the angle.

(9) $\frac{z_1}{z_2} = \frac{x_1+iy_1}{x_2+iy_2} = \frac{x_1+iy_1}{x_2+iy_2}\frac{x_2-iy_2}{x_2-iy_2} = \frac{x_1x_2-ix_1y_2+ix_2y_1-i^2y_1y_2}{x_2^2+y_2^2} = \frac{x_1x_2+y_1y_2+i(x_2y_1-x_1y_2)}{x_2^2+y_2^2}$

(10) $z_1z_2 = (x_1+iy_1)(x_2+iy_2) = x_1x_2 + ix_1y_2 + ix_2y_1 + i^2y_1y_2$
$= x_1x_2 + ix_1y_2 + ix_2y_1 - y_1y_2$
$= |z_1|\cos\theta_1|z_2|\cos\theta_2 + i|z_1|\cos\theta_1|z_2|\sin\theta_2 + i|z_1|\sin\theta_1|z_2|\cos\theta_2 - |z_1|\sin\theta_1|z_2|\sin\theta_2$
$= |z_1||z_2|(\cos\theta_1\cos\theta_2 - \sin\theta_1\sin\theta_2 + \cos\theta_1\sin\theta_2 + \sin\theta_1\cos\theta_2)$
$= |z_1||z_2|[\cos(\theta_1+\theta_2) + i\sin(\theta_1+\theta_2)]$

(11) $\bar{z}z = (x-iy)(x+iy) = x^2 + ixy - ixy - i^2y^2 = x^2 + y^2 = |z|^2$

Now divide by z on both sides to get $\bar{z} = \frac{|z|^2}{z}$.

(12) $z_1 + z_2 = x_1 + iy_1 + x_2 + iy_2 = x_1 + x_2 + i(y_1+y_2) \rightarrow \overline{z_1+z_2} = x_1 + x_2 - i(y_1+y_2)$
$\bar{z_1} + \bar{z_2} = x_1 - iy_1 + x_2 - iy_2 = x_1 + x_2 - i(y_1+y_2) = \overline{z_1+z_2}$

(13) $z_1z_2 = (x_1+iy_1)(x_2+iy_2) = x_1x_2 + ix_1y_2 + ix_2y_1 + i^2y_1y_2$
$= x_1x_2 + ix_1y_2 + ix_2y_1 - y_1y_2 = x_1x_2 - y_1y_2 + i(x_1y_2 + y_1x_2)$
$\overline{z_1z_2} = x_1x_2 - y_1y_2 - i(x_1y_2 + y_1x_2)$
$\bar{z_1}\,\bar{z_2} = (x_1-iy_1)(x_2-iy_2) = x_1x_2 - ix_1y_2 - ix_2y_1 + i^2y_1y_2$
$= x_1x_2 - ix_1y_2 - ix_2y_1 - y_1y_2 = x_1x_2 - y_1y_2 - i(x_1y_2 + y_1x_2) = \overline{z_1z_2}$

(14) $\frac{z-\bar{z}}{2i} = \frac{x+iy-(x-iy)}{2i} = \frac{x+iy-x+iy}{2i} = \frac{2iy}{2i} = y = \text{Im}(z)$

(15) $z_1z_2 = (x_1+iy_1)(x_2+iy_2) = x_1x_2 + ix_1y_2 + ix_2y_1 + i^2y_1y_2$
$= x_1x_2 + ix_1y_2 + ix_2y_1 - y_1y_2 = x_1x_2 - y_1y_2 + i(x_1y_2 + y_1x_2)$
$|z_1z_2| = \sqrt{(x_1x_2-y_1y_2)^2 + (x_1y_2+y_1x_2)^2}$
$|z_1z_2| = \sqrt{x_1^2x_2^2 - 2x_1x_2y_1y_2 + y_1^2y_2^2 + x_1^2y_2^2 + 2x_1y_2y_1x_2 + x_2^2y_1^2}$
$|z_1z_2| = \sqrt{x_1^2x_2^2 + y_1^2y_2^2 + x_1^2y_2^2 + x_2^2y_1^2}$
$|z_1||z_2| = \sqrt{x_1^2+y_1^2}\sqrt{x_2^2+y_2^2} = \sqrt{(x_1^2+y_1^2)(x_2^2+y_2^2)}$
$= \sqrt{x_1^2x_2^2 + x_1^2y_2^2 + x_2^2y_1^2 + y_1^2y_2^2} = |z_1z_2|$

Answer Key

(16) $z^{-1} = \frac{1}{z} = \frac{1}{x+iy} = \frac{1}{x+iy}\frac{x-iy}{x-iy} = \frac{x-iy}{x^2+y^2}$

(17) The problem is that all possible roots are not included. Since $(-1)^2$ and 1^2 both equal 1, there are two solutions for the square root of one: $1^{1/2} = \sqrt{1} = \pm 1$. Similarly, since $(-i)^2 = (-1)^2 i^2 = (1)(-1) = -1$ and $i^2 = -1$, there are two solutions to the square root of minus one: $(-1)^{1/2} = \sqrt{-1} = \pm i$. When these roots are considered independently, there are four combinations (two for the numerator and two for the denominator, where the signs need not match). The ratio is thus $\pm i$.

(18) Allowing the full periodicity, $e^{i\pi + 2i\pi k} = -1$ and $\ln(-1) = i\pi + 2\pi i k$ (as shown in Example 1). Thus, $\ln(-1) + \ln(-1) = i\pi + 2\pi i k_1 + i\pi + 2\pi i k_2 = 2i\pi + 2\pi i(k_1 + k_2)$ (allowing for different integers for each logarithm) is the general answer. For the special case of $k_1 = k_2 = 0$, we get $2i\pi$, and for the special case of $k_1 = 0$ and $k_2 = -1$ (or $k_1 = -1$ and $k_2 = 0$), we get 0.

(19) A cube root has three solutions.

$(-i)^{1/3} = \left(e^{-i\pi/2}\right)^{1/3} = \sqrt[3]{|1|}\left[\cos\left(\frac{-\pi/2 + 2\pi k}{3}\right) + i\sin\left(\frac{-\pi/2 + 2\pi k}{3}\right)\right]$

$(-i)^{1/3} = \sqrt[3]{|1|}\left[\cos\left(-\frac{\pi}{6} + \frac{2\pi k}{3}\right) + i\sin\left(-\frac{\pi}{6} + \frac{2\pi k}{3}\right)\right]$

One root is $(-i)^{1/3} = \cos\left(-\frac{\pi}{6}\right) + i\sin\left(-\frac{\pi}{6}\right) = \frac{\sqrt{3}}{2} - \frac{i}{2}$.

Since $-\frac{\pi}{6} + \frac{2\pi}{3} = \frac{\pi}{2}$, another root is $(-i)^{1/3} = \cos\frac{\pi}{2} + i\sin\frac{\pi}{2} = 0 - i = -i$.

Since $-\frac{\pi}{6} + \frac{4\pi}{3} = \frac{7\pi}{6}$, another root is $(-i)^{1/3} = \cos\frac{7\pi}{6} + i\sin\frac{7\pi}{6} = -\frac{\sqrt{3}}{2} - \frac{i}{2}$.

When we find all three roots, the three roots of $\left(e^{-i\pi/2}\right)^{1/3}$ are the same as the three roots of $\left(e^{3i\pi/2}\right)^{1/3}$. When you only find one root for each, funny things can happen.

(20) $1^x = 1$ if x is real, but what if the exponent is complex?

For complex exponentiation, $z^w = e^{w\ln z} = e^{w(i\theta + 2\pi i k + \ln|z|)}$.

Let $z = 1$ and $w = 2i\pi$ to get $1^{2i\pi} = e^{2i\pi(i0 + 2\pi i k + \ln 1)} = e^{2i\pi(0i + 2\pi i k + 0)} = e^{4\pi^2 i^2 k} = e^{-4\pi^2 k}$. This only equals 1 for the special case $k = 0$.

The expression $\left(e^{2i\pi}\right)^{2i\pi}$ similarly ignores the full periodicity of complex exponentiation; it should be $\left(e^{1+2i\pi k}\right)^{1+2i\pi k}$, similar to the second example of a seeming paradox given prior to Example 1.

Chapter 17

(1) Try using the identity $\cos\theta = \dfrac{e^{i\theta}+e^{-i\theta}}{2}$. Replace θ with 2θ or 4θ to see that

$\cos(2\theta) = \dfrac{e^{2i\theta}+e^{-2i\theta}}{2}$ and $\cos(4\theta) = \dfrac{e^{4i\theta}+e^{-4i\theta}}{2}$ (we'll use these at the end).

$\cos^2\theta = \left(\dfrac{e^{i\theta}+e^{-i\theta}}{2}\right)^2 = \dfrac{e^{2i\theta}+2e^{i\theta}e^{-i\theta}+e^{-2i\theta}}{4} = \dfrac{e^{2i\theta}+2+e^{-2i\theta}}{4}$

$\cos^4\theta = \left(\dfrac{e^{2i\theta}+2+e^{-2i\theta}}{4}\right)^2 = \dfrac{e^{4i\theta}+4e^{2i\theta}+2e^{2i\theta}e^{-2i\theta}+4+4e^{-2i\theta}+e^{-4i\theta}}{16}$

$\cos^4\theta = \dfrac{e^{4i\theta}+4e^{2i\theta}+2+4+4e^{-2i\theta}+e^{-4i\theta}}{16} = \dfrac{e^{4i\theta}+4e^{2i\theta}+6+4e^{-2i\theta}+e^{-4i\theta}}{16}$

$\cos^4\theta = \dfrac{e^{4i\theta}+e^{-4i\theta}}{16} + \dfrac{e^{2i\theta}+e^{-2i\theta}}{4} + \dfrac{3}{8} = \dfrac{\cos(4\theta)}{8} + \dfrac{\cos(2\theta)}{2} + \dfrac{3}{8} = \dfrac{\cos(4\theta)+4\cos(2\theta)+3}{8}$

(2) Use the identity $\sin\theta = \dfrac{e^{i\theta}-e^{-i\theta}}{2i}$. Recall from the solution to Exercise 1 that

$\cos(2\theta) = \dfrac{e^{2i\theta}+e^{-2i\theta}}{2}$ and $\cos(4\theta) = \dfrac{e^{4i\theta}+e^{-4i\theta}}{2}$ (we'll use these at the end).

$\sin^2\theta = \left(\dfrac{e^{i\theta}-e^{-i\theta}}{2i}\right)^2 = \dfrac{e^{2i\theta}-2e^{i\theta}e^{-i\theta}+e^{-2i\theta}}{-4} = \dfrac{-e^{2i\theta}+2-e^{-2i\theta}}{4}$

$\sin^4\theta = \left(\dfrac{-e^{2i\theta}+2-e^{-2i\theta}}{4}\right)^2 = \dfrac{e^{4i\theta}-4e^{2i\theta}+2e^{2i\theta}e^{-2i\theta}+4-4e^{-2i\theta}+e^{-4i\theta}}{16}$

$\sin^4\theta = \dfrac{e^{4i\theta}-4e^{2i\theta}+2+4-4e^{-2i\theta}+e^{-4i\theta}}{16} = \dfrac{e^{4i\theta}-4e^{2i\theta}+6-4e^{-2i\theta}+e^{-4i\theta}}{16}$

$\sin^4\theta = \dfrac{e^{4i\theta}+e^{-4i\theta}}{16} - \dfrac{e^{2i\theta}+e^{-2i\theta}}{4} + \dfrac{3}{8} = \dfrac{\cos(4\theta)}{8} - \dfrac{\cos(2\theta)}{2} + \dfrac{3}{8} = \dfrac{\cos(4\theta)-4\cos(2\theta)+3}{8}$

(3) Apply the identities $\sin\theta = \dfrac{e^{i\theta}-e^{-i\theta}}{2i}$ and $\cos\theta = \dfrac{e^{i\theta}+e^{-i\theta}}{2}$.

$2\sin\left(\dfrac{\alpha+\beta}{2}\right)\cos\left(\dfrac{\alpha-\beta}{2}\right) = 2\left(\dfrac{e^{i(\alpha+\beta)/2}-e^{-i(\alpha+\beta)/2}}{2i}\right)\left(\dfrac{e^{i(\alpha-\beta)/2}+e^{-i(\alpha-\beta)/2}}{2}\right)$

$= 2\left(\dfrac{e^{i\alpha/2}e^{i\beta/2}-e^{-i\alpha/2}e^{-i\beta/2}}{2i}\right)\left(\dfrac{e^{i\alpha/2}e^{-i\beta/2}+e^{-i\alpha/2}e^{i\beta/2}}{2}\right)$

$= \dfrac{e^{i\alpha/2}e^{i\beta/2}e^{i\alpha/2}e^{-i\beta/2}+e^{i\alpha/2}e^{i\beta/2}e^{-i\alpha/2}e^{i\beta/2}-e^{-i\alpha/2}e^{-i\beta/2}e^{i\alpha/2}e^{-i\beta/2}-e^{-i\alpha/2}e^{-i\beta/2}e^{-i\alpha/2}e^{i\beta/2}}{2i}$

Note that $e^{i\alpha/2}e^{-i\alpha/2} = e^{i\beta/2}e^{-i\beta/2} = e^0 = 1$, $e^{i\alpha/2}e^{i\alpha/2} = e^{i\alpha}$, and $e^{i\beta/2}e^{i\beta/2} = e^{i\beta}$.

$2\sin\left(\dfrac{\alpha+\beta}{2}\right)\cos\left(\dfrac{\alpha-\beta}{2}\right) = \dfrac{e^{i\alpha}+e^{i\beta}-e^{-i\beta}-e^{-i\alpha}}{2i} = \sin\alpha + \sin\beta$

This identity can be tricky to prove without using complex numbers. For example, see the author's *Trig Identities Practice Workbook*.

Answer Key

(4) Apply the identity $\cos\theta = \frac{e^{i\theta}+e^{-i\theta}}{2}$.

$$2\cos\left(\frac{\alpha+\beta}{2}\right)\cos\left(\frac{\alpha-\beta}{2}\right) = 2\left(\frac{e^{i(\alpha+\beta)/2}+e^{-i(\alpha+\beta)/2}}{2}\right)\left(\frac{e^{i(\alpha-\beta)/2}+e^{-i(\alpha-\beta)/2}}{2}\right)$$

$$= 2\left(\frac{e^{i\alpha/2}e^{i\beta/2}+e^{-i\alpha/2}e^{-i\beta/2}}{2}\right)\left(\frac{e^{i\alpha/2}e^{-i\beta/2}+e^{-i\alpha/2}e^{i\beta/2}}{2}\right)$$

$$= \frac{e^{i\alpha/2}e^{i\beta/2}e^{i\alpha/2}e^{-i\beta/2}+e^{i\alpha/2}e^{i\beta/2}e^{-i\alpha/2}e^{i\beta/2}+e^{-i\alpha/2}e^{-i\beta/2}e^{i\alpha/2}e^{-i\beta/2}+e^{-i\alpha/2}e^{-i\beta/2}e^{-i\alpha/2}e^{i\beta/2}}{2}$$

Note that $e^{i\alpha/2}e^{-i\alpha/2} = e^{i\beta/2}e^{-i\beta/2} = e^0 = 1$, $e^{i\alpha/2}e^{i\alpha/2} = e^{i\alpha}$, and $e^{i\beta/2}e^{i\beta/2} = e^{i\beta}$.

$$2\cos\left(\frac{\alpha+\beta}{2}\right)\cos\left(\frac{\alpha-\beta}{2}\right) = \frac{e^{i\alpha}+e^{i\beta}+e^{-i\beta}+e^{-i\alpha}}{2} = \cos\alpha + \cos\beta$$

(5) $\lambda = 2+i$ or $\lambda = 2-i$

Identify $q = 1, r = \sqrt{2}, s = -\sqrt{2}$, and $t = 3$. Use the formula $(q-\lambda)(t-\lambda) - rs = 0$.

$$(1-\lambda)(3-\lambda) - \sqrt{2}(-\sqrt{2}) = 0$$
$$3 - \lambda - 3\lambda + \lambda^2 + 2 = 0$$
$$\lambda^2 - 4\lambda + 5 = 0$$

This is a quadratic equation with $a = 1, b = -4$, and $c = 5$.

$$\lambda = \frac{-b \pm \sqrt{b^2-4ac}}{2a} = \frac{-(-4)\pm\sqrt{(-4)^2-4(1)(5)}}{2(1)} = \frac{4\pm\sqrt{16-20}}{2}$$

$$\lambda = \frac{4\pm\sqrt{-4}}{2} = \frac{4\pm\sqrt{(4)(-1)}}{2} = \frac{4\pm 2i}{2} = 2\pm i$$

(6) $\lambda = 1$ or $\lambda = -1$

Identify $q = t = 0, r = -i$, and $s = i$. Use the formula $(q-\lambda)(t-\lambda) - rs = 0$.

$$(0-\lambda)(0-\lambda) - (-i)(i) = 0$$
$$\lambda^2 + i^2 = 0$$
$$\lambda^2 - 1 = 0$$
$$\lambda^2 = 1$$
$$\lambda = \pm 1$$

Appendix

(1) $z = 3$, $z = 6$, or $z = 12$

$z^3 - 21z^2 + 126z - 216 = 0$, $a = -21$, $b = 126$, $c = -216$

$Q = \frac{3b-a^2}{9} = \frac{3(126)-(-21)^2}{9} = \frac{378-441}{9} = -\frac{63}{9} = -7$

$R = \frac{9ab-27c-2a^3}{54} = \frac{9(-21)(126)-27(-216)-2(-21)^3}{54} = \frac{-23{,}814+5832-2(-9261)}{54} =$

$\frac{-17{,}982+18{,}522}{54} = \frac{540}{54} = 10$

$Q^3 + R^2 = (-7)^3 + 10^2 = -343 + 100 = -243 < 0 \to 3$ distinct, real roots

$\sqrt{Q^3 + R^2} = \sqrt{-243} = \sqrt{(-1)(81)(3)} = 9i\sqrt{3}$ (follow Example 1)

$S = \sqrt[3]{R + \sqrt{Q^3 + R^2}} = \sqrt[3]{10 + 9i\sqrt{3}}$, $T = \sqrt[3]{R - \sqrt{Q^3 + R^2}} = \sqrt[3]{10 - 9i\sqrt{3}}$

modulus $= \sqrt{10^2 + \left(9\sqrt{3}\right)^2} = \sqrt{100 + 81(3)} = \sqrt{100 + 243} = \sqrt{343} = \sqrt{7(49)} = 7\sqrt{7}$

argument $= \tan^{-1}\left(\frac{9\sqrt{3}}{10}\right) \approx \tan^{-1}(1.5429) \approx 1$ rad (Quad. I)

$S = \sqrt[3]{10 + 9i\sqrt{3}} \approx \sqrt[3]{7\sqrt{7}e^i} \approx \left(7\sqrt{7}e^i\right)^{1/3} \approx \left(7\sqrt{7}\right)^{1/3} e^{i/3} \approx \left(7\sqrt{7}\right)^{1/3} \cos 0.3333 +$

$\left(7\sqrt{7}\right)^{1/3} i \sin 0.3333 \approx 2.6458(0.945) + 2.6458(0.3272)i \approx 2.5 + 0.8657i$

$T \approx 2.5 - 0.8657i$ (see Example 1)

$S + T \approx 2.5 + 0.8657i + 2.5 - 0.8657i = 5$

$-\frac{S+T}{2} \approx -\frac{2.5+0.8657i+2.5-0.8657i}{2} = -2.5$

$\frac{\sqrt{3}}{2}i(S - T) \approx \sqrt{3}\frac{2.5+0.8657i-2.5+0.8657i}{2}i \approx \sqrt{3}(0.8657)i^2 = \sqrt{3}(0.8657)(-1) \approx -1.50$

$z_1 = S + T - \frac{a}{3} \approx 5 - \frac{(-21)}{3} = 5 + 7 = 12$

$z_2 = -\frac{S+T}{2} - \frac{a}{3} + \frac{\sqrt{3}}{2}i(S - T) = -2.5 - \frac{(-21)}{3} - 1.50 = -4 + 7 = 3$

$z_3 = -\frac{S+T}{2} - \frac{a}{3} - \frac{\sqrt{3}}{2}i(S - T) = -2.5 - \frac{(-21)}{3} + 1.50 = -1 + 7 = 6$

Check: $2z^3 + 252z = 2(3)^3 + 252(3) = 2(27) + 756 = 54 + 756 = 810$

agrees with $42z^2 + 432 = 42(3)^2 + 432 = 42(9) + 432 = 378 + 432 = 810$

$2z^3 + 252z = 2(6)^3 + 252(6) = 2(216) + 1512 = 432 + 1512 = 1944$

agrees with $42z^2 + 432 = 42(6)^2 + 432 = 42(36) + 432 = 1512 + 432 = 1944$

# Answer Key

$2z^3 + 252z = 2(12)^3 + 252(12) = 2(1728) + 3024 = 3456 + 3024 = 6480$

agrees with $42z^2 + 432 = 42(12)^2 + 432 = 42(144) + 432 = 6048 + 432 = 6480$

(2) $z = -2, z = 4 + i$, or $z = 4 - i$

$z^3 - 6z^2 + z + 34 = 0, a = -6, b = 1, c = 34$

$Q = \frac{3b-a^2}{9} = \frac{3(1)-(-6)^2}{9} = \frac{3-36}{9} = -\frac{33}{9} = -\frac{11}{3}$

$R = \frac{9ab-27c-2a^3}{54} = \frac{9(-6)(1)-27(34)-2(-6)^3}{54} = \frac{-54-918-2(-216)}{54} = \frac{-972+432}{54} = -\frac{540}{54} = -10$

$Q^3 + R^2 = \left(-\frac{11}{3}\right)^3 + (-10)^2 = -\frac{1331}{27} + 100 = -\frac{1331}{27} + \frac{2700}{27} = \frac{1369}{27} > 0$

→ 1 real root and 2 complex roots (follow Example 2)

$\sqrt{Q^3 + R^2} = \sqrt{\frac{1369}{27}} = \frac{\sqrt{1369}}{\sqrt{27}} = \frac{37}{\sqrt{(9)(3)}} = \frac{37}{3\sqrt{3}} = \frac{37\sqrt{3}}{3\sqrt{3}\sqrt{3}} = \frac{37\sqrt{3}}{3(3)} = \frac{37\sqrt{3}}{9} \approx 7.12$

Note: $\frac{37\sqrt{3}}{9}$ is the 'proper' answer, but unnecessary since this isn't a final answer. It would be efficient to type sqrt(1369/27) into your calculator and get 7.12 in one step.

$S = \sqrt[3]{R + \sqrt{Q^3 + R^2}} \approx \sqrt[3]{-10 + 7.12} \approx \sqrt[3]{-2.88} \approx -1.42$

$T \approx \sqrt[3]{R - \sqrt{Q^3 + R^2}} \approx \sqrt[3]{-10 - 7.12} \approx \sqrt[3]{-17.12} \approx -2.58$

$z_1 = S + T - \frac{a}{3} \approx -1.42 - 2.58 - \frac{(-6)}{3} = -4 + 2 = -2$

$z_2 = -\frac{S+T}{2} - \frac{a}{3} + \frac{\sqrt{3}}{2}i(S - T) \approx -\frac{-1.42-2.58}{2} - \frac{(-6)}{3} + \frac{\sqrt{3}}{2}i[-1.42 - (-2.58)]$

$z_2 \approx 2 + 2 + \frac{1.16\sqrt{3}}{2}i \approx 4 + 1i = 4 + i$

$z_3 \approx 4 - i$

Check: $5z^3 + 5z + 170 = 5(-2)^3 + 5(-2) + 170 = 5(-8) - 10 + 170 = -40 + 160$
$= 120$ agrees with $30z^2 = 30(-2)^2 = 30(4) = 120$

$5z^3 + 5z + 170 = 5(4+i)^3 + 5(4+i) + 170 = 5(4+i)(16 + 8i - 1) + 20 + 5i + 170$
$= 5(64 + 32i - 4 + 16i - 8 - i) + 190 + 5i = 5(52 + 47i) + 190 + 5i = 260 + 235i$
$+ 190 + 5i = 450 + 240i$ agrees with $30z^2 = 30(4+i)^2 = 30(16 + 8i - 1)$
$= 30(15 + 8i) = 450 + 240i$

(3) $z = -3$ or $z = 9$ (and 9 is a double root)

$z^3 - 15z^2 + 27z + 243 = 0, a = -15, b = 27, c = 243$

214

$Q = \frac{3b-a^2}{9} = \frac{3(27)-(-15)^2}{9} = \frac{81-225}{9} = -\frac{144}{9} = -16$

$R = \frac{9ab-27c-2a^3}{54} = \frac{9(-15)(27)-27(243)-2(-15)^3}{54} = \frac{-3645-6561-2(-3375)}{54} = \frac{-10{,}205+6750}{54} = -\frac{3455}{54} \approx -64$

$Q^3 + R^2 \approx (-16)^3 + (-64)^2 = -4096 + 4096 = 0 \to$ real roots & at least 2 are equal

$S = \sqrt[3]{R + \sqrt{Q^3 + R^2}} \approx \sqrt[3]{-64 + 0} = \sqrt[3]{-64} = -4$

$T \approx \sqrt[3]{R - \sqrt{Q^3 + R^2}} \approx \sqrt[3]{-64 - 0} = \sqrt[3]{-64} = -4$

$z_1 = S + T - \frac{a}{3} \approx -4 - 4 - \frac{(-15)}{3} = -8 + 5 = -3$

$z_2 = -\frac{S+T}{2} - \frac{a}{3} + \frac{\sqrt{3}}{2}i(S - T) \approx -\frac{-4-4}{2} - \frac{(-15)}{3} + \frac{\sqrt{3}}{2}i[-4 - (-4)]$

$z_2 \approx 4 + 5 + \frac{\sqrt{3}}{2}i(0) = 9$

$z_3 \approx 9$

Check: $\frac{z^3}{3} + 9z = \frac{(-3)^3}{3} + 9(-3) = -\frac{27}{3} - 27 = -9 - 27 = -36$

agrees with $5z^2 - 81 = 5(-3)^2 - 81 = 5(9) - 81 = 45 - 81 = -36$

$\frac{z^3}{3} + 9z = \frac{9^3}{3} + 9(9) = \frac{729}{3} + 81 = 243 + 81 = 324$

agrees with $5z^2 - 81 = 5(9)^2 - 81 = 5(81) - 81 = 405 - 81 = 324$

(4) $z = -6, z = \frac{2}{3}$, or $z = 2$ (and 2 is a double root)

$z^4 + \frac{4}{3}z^3 - \frac{64}{3}z^2 + \frac{112}{3}z - 16 = 0, e = \frac{4}{3}, f = -\frac{64}{3}, g = \frac{112}{3}, h = -16$

$a = -f = -\left(-\frac{64}{3}\right) = \frac{64}{3}$, $b = eg - 4h = \left(\frac{4}{3}\right)\left(\frac{112}{3}\right) - 4(-16) = \frac{448}{9} + 64 = \frac{1024}{9}$

$c = 4fh - g^2 - e^2h = 4\left(-\frac{64}{3}\right)(-16) - \left(\frac{112}{3}\right)^2 - \left(\frac{4}{3}\right)^2(-16)$

$c = \frac{4096}{3} - \frac{12{,}544}{9} + \frac{256}{9} = \frac{12{,}288}{9} - \frac{12{,}544}{9} + \frac{256}{9} = 0$

$y^3 + ay^2 + by + c = 0 \to y^3 + \frac{64}{3}y^2 + \frac{1024}{9}y = 0$

Since $c = 0$, this cubic is easier to solve than a general cubic. Since every term includes y, one of the three solutions is $y = 0$. We just need one real solution for y, and since zero is real, we don't need to bother with the other solutions: $y_r = 0$.

Answer Key

$$z^2 + \frac{e \pm \sqrt{e^2 - 4f + 4y_r}}{2}z + \frac{y_r \pm \sqrt{y_r^2 - 4h}}{2} = 0 \rightarrow z^2 + \frac{\frac{4}{3} \pm \sqrt{\left(\frac{4}{3}\right)^2 - 4\left(-\frac{64}{3}\right) + 4(0)}}{2}z + \frac{0 \pm \sqrt{0^2 - 4(-16)}}{2} = 0$$

$$z^2 + \frac{\frac{4}{3} \pm \sqrt{\frac{16}{9} + \frac{256}{3}}}{2}z \pm \frac{\sqrt{64}}{2} = z^2 + \frac{\frac{4}{3} \pm \sqrt{\frac{16}{9} + \frac{768}{9}}}{2}z \pm \frac{8}{2} = z^2 + \frac{\frac{4}{3} \pm \sqrt{\frac{784}{9}}}{2}z \pm 4 = z^2 + \frac{\frac{4}{3} \pm \frac{28}{3}}{2}z \pm 4 = 0$$

$$z^2 + \left(\frac{2}{3} \pm \frac{14}{3}\right)z \pm 4 = 0$$

The two combinations of signs that lead to correct answers for the quartic equation are $z^2 + \frac{16}{3}z - 4 = 0$ and $z^2 - \frac{12}{3}z + 4 = 0$. We will focus on this pair of quadratics.

$z^2 + \frac{16}{3}z - 4 = 0$ or $z^2 - 4z + 4 = 0$

$3z^2 + 16z - 12 = 0$ or $z^2 - 4z + 4 = 0$ (We multiplied the first equation by 3.)

$$z = \frac{-16 \pm \sqrt{16^2 - 4(3)(-12)}}{2(3)} = \frac{-16 \pm \sqrt{256 + 144}}{6} = \frac{-16 \pm \sqrt{400}}{6} = \frac{-16 \pm 20}{6}$$

or $z = \frac{-(-4) \pm \sqrt{(-4)^2 - 4(1)(4)}}{2(1)} = \frac{4 \pm \sqrt{16-16}}{2} = \frac{4 \pm \sqrt{0}}{2} = \frac{4}{2} = 2$ (a double root)

The first equation gives $z = \frac{-16-20}{6} = -\frac{36}{6} = -6$ or $z = \frac{-16+20}{6} = \frac{4}{6} = \frac{2}{3}$

The three solutions are $z = -6, z = \frac{2}{3}$, or $z = 2$.

Check: $3z^4 + 4z^3 + 112z = 3(-6)^4 + 4(-6)^3 + 112(-6)$
$= 3(1296) + 4(-216) - 672 = 3888 - 864 - 672 = 2352$ agrees with
$64z^2 + 48 = 64(-6)^2 + 48 = 64(36) + 48 = 2304 + 48 = 2352$

Check: $3z^4 + 4z^3 + 112z = 3(2)^4 + 4(2)^3 + 112(2) = 3(16) + 4(8) + 224$
$= 48 + 32 + 224 = 304$ agrees with $64z^2 + 48 = 64(2)^2 + 48 = 64(4) + 48 = 256 + 48 = 304$

Check: $3z^4 + 4z^3 + 112z = 3\left(\frac{2}{3}\right)^4 + 4\left(\frac{2}{3}\right)^3 + 112\left(\frac{2}{3}\right) = \frac{3(16)}{81} + \frac{4(8)}{27} + \frac{224}{3}$

$= \frac{16}{27} + \frac{32}{27} + \frac{2016}{27} = \frac{2064}{27} = \frac{688}{9}$ agrees with $64z^2 + 48 = 64\left(\frac{2}{3}\right)^2 + 48 = 64\left(\frac{4}{9}\right) + 48$

$= \frac{256}{9} + \frac{432}{9} = \frac{688}{9}$

(5) $z = -12, z = -2, z = 2$, or $z = 6$

$z^4 + 6z^3 - 76z^2 - 24z + 288 = 0, e = 6, f = -76, g = -24, h = 288$

$a = -f = -(-76) = 76$

$b = eg - 4h = (6)(-24) - 4(288) = -144 - 1152 = -1296$

$c = 4fh - g^2 - e^2h = 4(-76)(288) - (-24)^2 - (6)^2(288)$

$c = -87{,}552 - 576 - 10{,}368 = -98{,}496$

$y^3 + ay^2 + by + c = 0 \rightarrow y^3 + 76y^2 - 1296y - 98{,}496 = 0$

$Q = \frac{3b-a^2}{9} = \frac{3(-1296)-(76)^2}{9} = \frac{-3888-5776}{9} = -\frac{9664}{9} \approx -1074$

$R = \frac{9ab-27c-2a^3}{54} = \frac{9(76)(-1296)-27(-98{,}496)-2(76)^3}{54} = \frac{-886{,}464+2{,}659{,}392-2(438{,}976)}{54} = \frac{1{,}772{,}928-877{,}952}{54} = \frac{894{,}976}{54} = \frac{447{,}488}{27} \approx 16{,}574$

$Q^3 + R^2 \approx (-1074)^3 + 16{,}574^2 \approx -1.239 \times 10^9 + 2.747 \times 10^8 \approx -9.643 \times 10^8$

$Q < 0 \rightarrow$ 3 distinct, real roots (for the cubic, not for the quartic)

$\sqrt{Q^3 + R^2} \approx \sqrt{-9.643 \times 10^8} \approx 31{,}053i$ (follow Example 1)

$S = \sqrt[3]{R + \sqrt{Q^3 + R^2}} \approx \sqrt[3]{16{,}574 + 31{,}053i},\ T = \sqrt[3]{R - \sqrt{Q^3 + R^2}} \approx \sqrt[3]{16{,}574 - 31{,}053i}$

modulus $\approx \sqrt{16{,}574^2 + 31{,}053^2} \approx \sqrt{2.747 \times 10^8 + 9.643 \times 10^8} \approx \sqrt{1.239 \times 10^9} \approx 35{,}199$

argument $\approx \tan^{-1}\left(\frac{31{,}053}{16{,}574}\right) \approx \tan^{-1}(1.874) \approx 1.081$ rad (Quad. I)

$S \approx \sqrt[3]{16{,}574 + 31{,}053i} \approx \sqrt[3]{35{,}199 e^{1.081i}} \approx (35{,}199 e^{1.081i})^{1/3} \approx 35{,}199^{1/3} e^{1.081i/3}$
$\approx 32.77 e^{0.3603i} \approx 32.77 \cos 0.3603 + 32.77i \sin 0.3603 \approx 30.67 + 11.55i$

$T \approx 30.67 - 11.55i$ (see Example 1)

$S + T \approx 30.67 + 11.55i + 30.67 - 11.55i = 61.34$

$y_1 = S + T - \frac{a}{3} \approx 61.34 - \frac{76}{3} \approx 61.34 - 25.33 \approx 36$

We just need one real solution for the cubic, so we don't need to bother finding y_2 and y_3. Call this solution $y_r \approx 36$.

Tip: Plug the answer for y_r into the cubic equation to verify that correctly solves the cubic before proceeding to solve the quartic. (If you know that y_r is incorrect, you need to find your mistake before going further.)

Cubic check: $y^3 + 76y^2 - 1296y - 98{,}496 \approx 36^3 + 76(36)^2 - 1296(36) - 98{,}496 = 46{,}656 + 98{,}496 - 46{,}656 - 98{,}496 = 0$ (Breathe a sigh of relief.)

Recall that $e = 6, f = -76, g = -24$, and $h = 288$.

$z^2 + \frac{e \pm \sqrt{e^2 - 4f + 4y_r}}{2} z + \frac{y_r \pm \sqrt{y_r^2 - 4h}}{2} = 0 \rightarrow z^2 + \frac{6 \pm \sqrt{6^2 - 4(-76) + 4(36)}}{2} z + \frac{36 \pm \sqrt{36^2 - 4(288)}}{2} = 0$

Answer Key

$$z^2 + \frac{6\pm\sqrt{6^2-4(-76)+4(36)}}{2}z + \frac{36\pm\sqrt{36^2-4(288)}}{2} = 0 \to z^2 + \frac{6\pm\sqrt{36+304+144}}{2}z + \frac{36\pm\sqrt{1296-1152}}{2} = 0$$

$$z^2 + \frac{6\pm\sqrt{484}}{2}z + \frac{36\pm\sqrt{144}}{2} = 0 \to z^2 + \frac{6\pm 22}{2}z + \frac{36\pm 12}{2} = 0 \to z^2 + (3\pm 11)z + 18 \pm 6 = 0$$

The two combinations of signs that lead to correct answers for the quartic equation are $z^2 + (3+11)z + 18 + 6 = 0$ and $z^2 + (3-11)z + 18 - 6 = 0$. We will focus on this pair of quadratics.

$z^2 + 14z + 24 = 0$ or $z^2 - 8z + 12 = 0$

$$z = \frac{-14\pm\sqrt{14^2-4(1)(24)}}{2(1)} = \frac{-14\pm\sqrt{196-96}}{2} = \frac{-14\pm\sqrt{100}}{2} = \frac{-14\pm 10}{2} = -7 \pm 5$$

or $$z = \frac{-(-8)\pm\sqrt{(-8)^2-4(1)(12)}}{2(1)} = \frac{8\pm\sqrt{64-48}}{2} = \frac{8\pm\sqrt{16}}{2} = \frac{8\pm 4}{2} = 4 \pm 2$$

The first equation gives $z = -2$ or $z = -12$. The second equation gives $z = 2$ or $z = 6$.
Check: $288 - 76z^2 + z^4 = 288 - 76(\pm 2)^2 + (\pm 2)^4 = 288 - 76(4) + 16 = 0$ agrees with $24z - 6z^3 = 24(\pm 2) - 6(\pm 2)^3 = \pm 48 - 6(\pm 8) = \pm 48 \mp 48 = 0$.
$288 - 76z^2 + z^4 = 288 - 76(-12)^2 + (-12)^4 = 288 - 76(144) + 20{,}736 = 10{,}080$ agrees with $24z - 6z^3 = 24(-12) - 6(-12)^3 = -288 - 6(-1728) = -288 + 10{,}368 = 10{,}080$.
$288 - 76z^2 + z^4 = 288 - 76(6)^2 + (6)^4 = 288 - 76(36) + 1296 = -1152$ agrees with $24z - 6z^3 = 24(6) - 6(6)^3 = 144 - 1296 = -1152$.

(6) $z = 1 + \sqrt{3}, z = 1 - \sqrt{3}, z = 1 + i$, or $z = 1 - i$
$z^4 - 4z^3 + 4z^2 - 4 = 0, e = -4, f = 4, g = 0, h = -4$
Note: $g = 0$ because there isn't a term proportional to the first power of z.
$a = -f = -4$
$b = eg - 4h = (-4)(0) - 4(-4) = 0 + 16 = 16$
$c = 4fh - g^2 - e^2h = 4(4)(-4) - 0^2 - (-4)^2(-4)$
$c = -64 - 0 - (16)(-4) = -64 + 64 = 0$
$y^3 + ay^2 + by + c = 0 \to y^3 - 4y^2 + 16y + 0 = 0$
Since $c = 0$, this cubic is easier to solve than a general cubic. Since every term includes y, one of the three solutions is $y = 0$. We just need one real solution for y, and since zero is real, we don't need to bother with the other solutions: $y_r = 0$.

$$z^2 + \frac{e\pm\sqrt{e^2-4f+4y_r}}{2}z + \frac{y_r\mp\sqrt{y_r^2-4h}}{2} = 0 \to z^2 + \frac{-4\pm\sqrt{(-4)^2-4(4)+4(0)}}{2}z + \frac{0\mp\sqrt{0^2-4(-4)}}{2} = 0$$

Complex Numbers Essentials Math Workbook with Answers

$$z^2 + \frac{-4\pm\sqrt{16-16+0}}{2}z \pm \frac{\sqrt{16}}{2} = z^2 + \frac{-4\pm\sqrt{0}}{2}z \pm \frac{4}{2} = z^2 + \frac{-4}{2}z \pm 2 = z^2 - 2z \pm 2 = 0$$

Since the first $\pm$ sign preceded zero, and since $+0$ and -0 are equivalent, only the second $\pm$ sign is relevant. There are only 2 combinations to choose from in this problem, and we need them both to obtain all four roots. The two combinations are $z^2 - 2z - 2 = 0$ and $z^2 - 2z + 2 = 0$.

$$z = \frac{-(-2)\pm\sqrt{(-2)^2-4(1)(-2)}}{2(1)} = \frac{2\pm\sqrt{4+8}}{2} = \frac{2\pm\sqrt{12}}{2} = \frac{2\pm\sqrt{4(3)}}{2} = \frac{2\pm 2\sqrt{3}}{2} = 1 \pm \sqrt{3}$$

or $z = \frac{-(-2)\pm\sqrt{(-2)^2-4(1)(2)}}{2(1)} = \frac{2\pm\sqrt{4-8}}{2} = \frac{2\pm\sqrt{-4}}{2} = \frac{2\pm\sqrt{4(-1)}}{2} = \frac{2\pm 2i}{2} = 1 \pm i$

The four roots are $z = 1 - \sqrt{3}$, $z = 1 + \sqrt{3}$, $z = 1 - i$, and $z = 1 + i$.

Check: $(1 + \sqrt{3})^2 = 1 + 2\sqrt{3} + 3 = 4 + 2\sqrt{3}$

$(1 + \sqrt{3})^3 = (1 + \sqrt{3})^2(1 + \sqrt{3}) = (4 + 2\sqrt{3})(1 + \sqrt{3}) = 4 + 4\sqrt{3} + 2\sqrt{3} + 2(3)$
$= 4 + 6\sqrt{3} + 6 = 10 + 6\sqrt{3}$

$(1 + \sqrt{3})^4 = (1 + \sqrt{3})^3(1 + \sqrt{3}) = (10 + 6\sqrt{3})(1 + \sqrt{3}) = 10 + 10\sqrt{3} + 6\sqrt{3} + 6(3)$
$= 10 + 16\sqrt{3} + 18 = 28 + 16\sqrt{3}$

$\frac{z^4}{4} - z^3 + z^2 = \frac{(1+\sqrt{3})^4}{4} - (1 + \sqrt{3})^3 + (1 + \sqrt{3})^2 = \frac{28+16\sqrt{3}}{4} - (10 + 6\sqrt{3}) + 4 + 2\sqrt{3}$
$= 7 + 4\sqrt{3} - 10 - 6\sqrt{3} + 4 + 2\sqrt{3} = 1 + 0\sqrt{3} = 1$

$(1 + i)^2 = 1 + 2i + i^2 = 1 + 2i - 1 = 2i$

$(1 + i)^3 = (1 + i)^2(1 + i) = 2i(1 + i) = 2i + 2i^2 = 2i - 2$

$(1 + i)^4 = (1 + i)^3(1 + i) = (2i - 2)(1 + i) = 2i + 2i^2 - 2 - 2i = -2 - 2 = -4$

$\frac{z^4}{4} - z^3 + z^2 = \frac{(1+i)^4}{4} - (1 + i)^3 + (1 + i)^2 = \frac{-4}{4} - (2i - 2) + 2i = -1 - 2i - (-2) + 2i$
$= -1 - 2i + 2 + 2i = 1 + 0i = 1$

We know from Chapter 15 that if the root of a polynomial has the form $p + qi$ or the form $t + \sqrt{u}$ and if the coefficients of the polynomial are rational, then the conjugate of that root is also a root, so we don't need to check $z = 1 - \sqrt{3}$ or $z = 1 - i$ to know that they are also correct solutions.

Index

A

Abel 114 (fn 1)
AC circuits 28, 126-127
addition (complex) 20-26, 111
additive inverse 113
adjacent 63
alternate solutions 76-77, 89
angle sum identity 9-10, 65, 113
appendix 133-146
applications 9-10, 123-132
Argand diagram 27-30
argument 35-36, 57, 111
arithmetic (complex) 20-26, 111
associative property 113
asterisk 17

C

calculus 65-66, 124
capacitor 126
Cartesian form 35, 38-39, 62, 111
Cauchy's integral formula 124
circuits 28, 126-127
Clausen 114 (fn 1)
coefficient 94
combine like terms 20
commutative property 113
complex angle 64-67
complex arithmetic 20-26, 111
complex conjugate 17-19, 58, 89, 98-99, 111

complex exponentiation 114

complex logarithm 67

complex numbers 14-16

complex numbers (properties of) 111-122

complex pairs 89, 99

complex plane 27-30, 79, 82

complex roots 41, 75-94, 98, 102, 133

complex trigonometry 63-74

conjugate square roots 99

constant 94

contour integral 124

contradictions 11, 114-115, 117-118, 122

cosh 64-67, 112-113

cosine 63-67, 112-113

cosine (hyperbolic) 63-67, 112-113

cube root 78

cubic formula 133-146

cubic polynomial 96, 133-146

cubic term 94

current 126

D

degree (of a polynomial) 94

degrees 36, 57

de Moivre's theorem 49-56, 58, 112

denominator 100

determinant 124-125

discriminant (cubic) 133

discriminant (quadratic) 41, 124

distinct roots 94, 133

distribute 20-21

distributive property 113

Index

dividend 97

division (complex) 21-26, 111

division (polynomial) 97

divisor 97

domain error 67

double root 41, 94, 133

E

eigenvalues 124-125

electromagnetic field 125

Euler's formula 57-62, 66, 112

Euler's number 57

even function 65

exponential form 57, 111, 114

exponential function 57-58, 64, 66-67, 112, 114

exponentiation (complex) 114

exponent of a complex number 51-56, 114

exponent rules 114

F

factor 20, 32, 41, 90, 98

factorial 66

factoring polynomials 90, 94-110, 123

factoring the quadratic 41-48

factorization (polynomial) 90, 94-95, 97-102

factor theorem 98

Fourier transform 127

fraction 6, 99-101

fundamental theorem of algebra 98, 123

G

geometric series 90

graph (polynomial) 102

H

Hilbert space 126

Hippasus' proof 6 (fn 1)

hyperbolic functions 64-67, 112-113

hypotenuse 63

I

identities (complex numbers) 111-122

identities (trig) 9-10, 63-67, 74, 112-113

identity matrix 124

i (imaginary number) 8-13

imaginary angle 64

imaginary axis 27, 36

imaginary numbers 5-13, 63, 67

imaginary part 14, 40, 58, 112

imaginary (purely) 14-16, 27, 36

imaginary roots 41, 94

imaginary solution 94

impedance 126

improper integral 124

induction (method of) 49-51

inductor 126

integers 99-101

inverse (additive) 113

inverse (hyperbolic) 67, 69

inverse (multiplicative) 113

inverse tangent 36, 67

Index

inverse (trig) 67, 69
irrational numbers 6

L

Lie algebra 125
linear algebra 124-125
linear polynomial 96
linear term 94
logarithm 67, 114

M

Maclaurin series 65-66
magnitude 31 (fn 1)
matrix algebra 124-125
matrix exponentiation 125
method of induction 49-51
mixed fraction 6
modulus 31-36, 57, 111-112
modulus-square 31
multiple roots 76-77, 89
multiplication (complex) 21-26, 111
multiplicative inverse 113
multiplicity 94
multi-valued 114

N

negative angle identity 113
negative numbers 5-6, 101
no solution 7
numerator 100

O

odd function 65
opposite 63
origin 27
overbar 17

P

pairs of roots 89, 99
paradoxes 11, 114-115, 117-118, 122
Pauli spin matrices 125, 132
percent 6
perfect square 32
periodic 67, 113-115
phase shift 126
phasor 28, 126
plane 27-30
polar form 35-38, 57, 111
polygon 79, 82, 89
polynomial division 97
polynomial factorization 90, 94-95
polynomial roots 94-110, 123
power of a complex number 51-56
power rules 114
powers of i 8-13, 111
principal value 115
properties of complex numbers 111-122
properties of exponentials 57-58
properties of trig functions 64-67
purely imaginary 14-16, 27, 36
purely real 14-16, 27, 36, 123
Pythagorean identity 63
Pythagorean theorem 31, 63

Index

Q

quadratic equation 41-48

quadratic formula 41, 96

quadratic polynomial 41-48, 96

quadratic term 94

quantum mechanics 125-127

quartic formula 133-146

quartic polynomial 96, 133-146

quartic term 94

quotient 97

R

radians 36, 57, 66, 76

rational numbers 5, 99

reactance 126-127

real axis 27-30, 36

real numbers 5, 7

real part 14, 39-40, 58, 112

real (purely) 14-16, 27, 36, 123

real roots 41, 133

real value 17-18

rectangular form 35, 111

reduced fraction 100

remainder (relating to powers of i) 9

remainder theorem 97

repeating decimal 6

residue theorem 124

resistance 126

resonance 127

roots (complex) 41, 75-94, 98, 102, 114, 133

roots (cube) 78

roots (imaginary) 41, 94

roots of polynomials 94-110, 123

roots of unity 89-93

roots (quadratic) 41-48

roots (real) 41, 133

roots (square root of a complex number) 51-56, 75-88

S

Schrödinger's equation 126

series 65-66, 90

signal processing 127

signs 102

sine 64-67, 112-113

sine (hyperbolic) 64-67, 112-113

sinh 64-67, 112-113

spin angular momentum 125

square root of a complex number 51-56, 75-88, 165-166

square root of i 75-76

square root of negative one 8, 11

square roots in polynomials 99, 102

standard form (cubic) 133-134

standard form (quadratic) 41

standard form (quartic) 134

Steiner 114 (fn 1)

subtraction (complex) 20-26, 111

T

tangent 36

tangent (hyperbolic) 65

Taylor series 66 (fn 1)

term (polynomial) 94

Index

transitive property 113

trig identities 9-10, 63-67, 74, 112-113, 123

trigonometry 54, 63-74, 112-113, 123

U

unity (roots of) 89-93

V

vector 28, 31 (fn 1)

Vieta's formulas 96

W

wave equation 127

wave function 126

whole numbers 5

Z

zero (number) 5

zero (of a polynomial) 94

zero to the power of zero 8 (fn 2)

WAS THIS BOOK HELPFUL?

Much effort and thought were put into this book, such as:
- Introducing the main ideas at the beginning of each chapter. The goal was to be concise while also covering the pertinent information.
- Solving examples step by step to serve as a helpful guide.
- Including not just answers, but even the solutions to most of the problems.
- Emphasizing in Chapter 1 and Chapter 17 that there are many practical applications of complex numbers. It's not merely a fascinating subject; it's actually put to good use.

If you appreciate the effort that went into making this book possible, there is a simple way that you could show it:

Please take a moment to post an honest review.

For example, you can review this book at Amazon.com or Goodreads.com.

Even a short review can be helpful and will be much appreciated. If you are not sure what to write, following are a few ideas, though it is best to describe what is important to you.
- Was it helpful to have notes/solutions in addition to the answers at the back of the book?
- Were the examples useful?
- Were you able to understand the ideas at the beginning of the chapter?
- Did this book offer good practice for you?
- Would you recommend this book to others? If so, why?

Do you believe that you found a mistake? Please email the author, Chris McMullen, at greekphysics@yahoo.com to ask about it. One of two things will happen:
- You might discover that it wasn't a mistake after all and learn why.
- You might be right, in which case the author will be grateful and future readers will benefit from the correction. Everyone is human.

ABOUT THE AUTHOR

Dr. Chris McMullen has over 20 years of experience teaching university physics in California, Oklahoma, Pennsylvania, and Louisiana. Dr. McMullen is also an author of math and science workbooks. Whether in the classroom or as a writer, Dr. McMullen loves sharing knowledge and the art of motivating and engaging students.

The author earned his Ph.D. in phenomenological high-energy physics (particle physics) from Oklahoma State University in 2002. Originally from California, Chris McMullen earned his Master's degree from California State University, Northridge, where his thesis was in the field of electron spin resonance.

As a physics teacher, Dr. McMullen observed that many students lack fluency in essential math skills. In an effort to help students of all ages and levels become fluent in mathematics, he published a series of math workbooks on fractions, long division, word problems, algebra, geometry, trigonometry, logarithms, calculus, probability, differential equations, and more. Dr. McMullen has also published a variety of science books, including astronomy, chemistry, and physics workbooks.

Author, Chris McMullen, Ph.D.

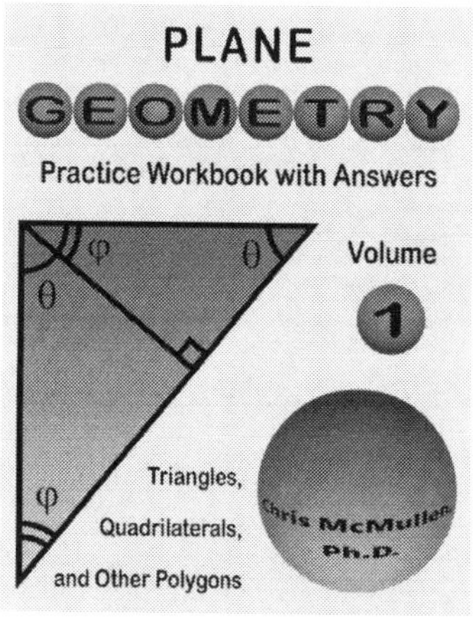
PLANE
GEOMETRY
Practice Workbook with Answers
Volume
1
Triangles,
Quadrilaterals,
and Other Polygons
Chris McMullen
Ph.D.

PLANE
GEOMETRY
Practice Workbook with Answers
Volume
2
Circles, Chords,
Secants, and Tangents
Chris McMullen
Ph.D.

GEOMETRY
PROOFS
ESSENTIAL PRACTICE PROBLEMS
WORKBOOK WITH FULL SOLUTIONS
CHRIS MCMULLEN, PH.D.

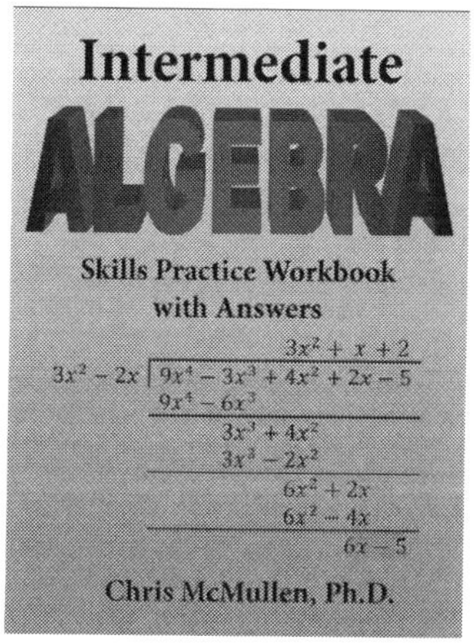
Intermediate
ALGEBRA
Skills Practice Workbook
with Answers
Chris McMullen, Ph.D.

101 Involved
ALGEBRA
Problems with Answers
Chris McMullen, Ph.D.

Master Essential
ALGEBRA
Skills Practice Workbook
with Answers
Chris McMullen, Ph.D.

Chris McMullen, Ph.D.
Essential Skills Practice Workbook with Answers
LOGARITHMS
and
EXPONENTIALS

TRIGONOMETRY
Essentials Practice Workbook with Answers
Master Basic Trig Skills
Improve Your Math Fluency Series
Chris McMullen, Ph.D.

TRIG
IDENTITIES
Practice Workbook with Answers
Chris McMullen, Ph.D.

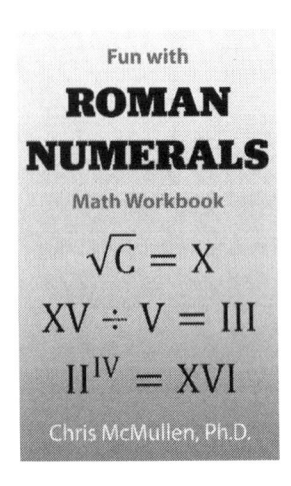

Fun with ROMAN NUMERALS
Math Workbook

$\sqrt{C} = X$

$XV \div V = III$

$II^{IV} = XVI$

Chris McMullen, Ph.D.

300+ MATHEMATICAL PATTERN PUZZLES
NUMBER PATTERN RECOGNITION AND REASONING

CHRIS MCMULLEN, PH.D.

PYRAMID MATH PUZZLE CHALLENGE
CAN YOU FILL IN THE MISSING NUMBERS?

CHRIS MCMULLEN, PH.D.

The FOUR-COLOR THEOREM and Basic GRAPH THEORY

Chris McMullen, Ph.D.

WORD PROBLEMS with Answers

Sam is 20 years older than Amy. 8 years ago, Sam was twice as old as Amy. How old are Sam and Amy now?

	8 years ago	now
Amy	20	28
Sam	40	48

Chris McMullen, Ph.D.

ESSENTIAL PROBABILITY PRACTICE WORKBOOK WITH ANSWERS

CHRIS MCMULLEN, PH.D.

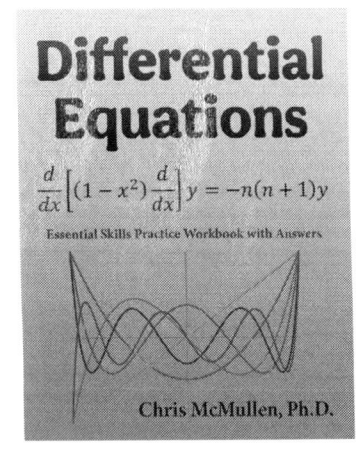

Differential Equations

$\frac{d}{dx}\left[(1-x^2)\frac{d}{dx}\right]y = -n(n+1)y$

Essential Skills Practice Workbook with Answers

Chris McMullen, Ph.D.

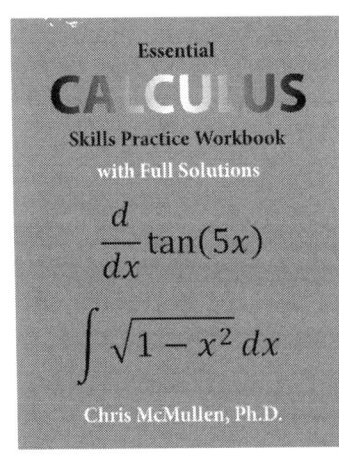

Essential CALCULUS
Skills Practice Workbook with Full Solutions

$\frac{d}{dx}\tan(5x)$

$\int \sqrt{1-x^2}\,dx$

Chris McMullen, Ph.D.

CALCULUS with Multiple Variables
Essential Skills Workbook

$\vec{\nabla}f$ $\vec{\nabla}\cdot\vec{A}$

$\frac{\partial^2}{\partial x \partial y}3\ln(x+y)$

Includes Vector Calculus and Full Solutions

$\int_{x=0}^{2}\int_{y=4}^{9} x\sqrt{y}\,dx\,dy$

$\vec{\nabla}\times\vec{A}$

Chris McMullen, Ph.D.

Printed in Great Britain
by Amazon